Contents

How to use this book

Each page has a title telling you what it is about.

Adding and subtracting

WN2.15a

Find the difference between the number of miles travelled by each car.

Hint: Use the nearest multiple of 10. You can choose whether to count back or count on.

This Hint will help you to answer the question.

This shows how to set out your work.

1

246 miles 59 miles

2
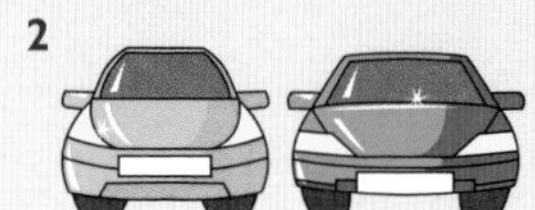
427 miles 79 miles

3
154 miles 88 miles

4

285 miles 79 miles

5
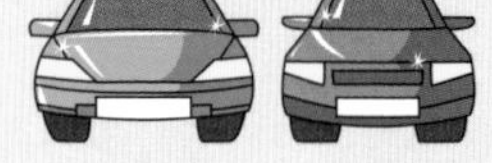
274 miles 57 miles

6

164 miles 89 miles

Instructions look like this. Always read these carefully before starting.

How many centimetres must each snail crawl to reach the lettuce?

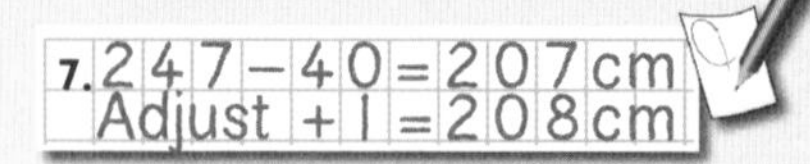

7: 39 cm 8: 67 cm 9: 99 cm 10: 128 cm 11: 189 cm lettuce: 247 cm

Work with a partner. Agree who will count back and who will count on. Try this subtraction: 167 – 89. Reverse roles. Which is the best method for you? Try 224 – 129.

These are Rocket activities. Ask your teacher if you need to do these questions.

I can do a calculation by rounding one of the numbers and then adjust my answer at the end.

Read this to check you understand what you have been learning on the page.

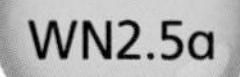

6-digit numbers

1 Write the number that matches each letter on the line.

1. a: 576400

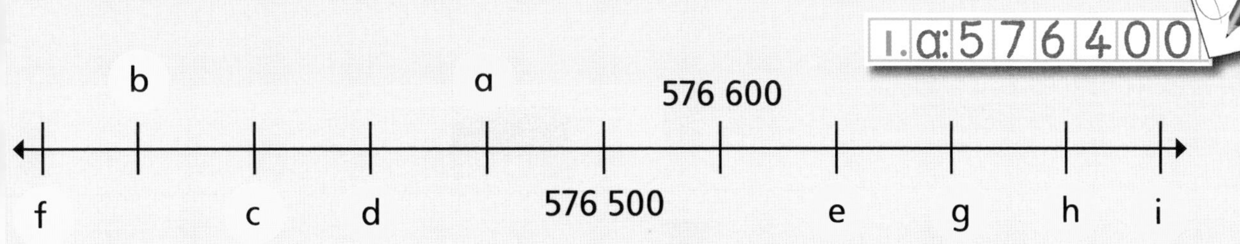

Write each pair of numbers in figures. Use < or > between them.

2 Fifty-six thousand, four hundred and two

Twenty-nine thousand, six hundred and seventeen

3 Sixty-four thousand, seven hundred and eighty-two

Seven hundred and thirty thousand, two hundred and twelve

4 Three hundred thousand, eight hundred and three

Fifty-four thousand, six hundred and sixty

5 Two hundred and seventy-eight thousand, nine hundred and twenty

Twelve thousand, seven hundred and eighty-six

6 One hundred and four thousand, two hundred and two

Ninety-nine thousand, nine hundred and ninety-nine

7 Thirty-seven thousand, four hundred and seventy-two

Eight hundred and twenty-seven thousand, four hundred and six

How many different numbers can be used to complete this number sentence: 10 998 < ☐ < 11 100?

I can write and order 6-digit numbers

6-digit numbers

Write the next four numbers in each sequence.

1. 59473, 60473, 61473, 62473

1 56473, 57473, 58473, . . .

2 24608, 24708, 24808, . . .

3 66419, 56419, 46419, . . .

4 162308, 172308, 182308, . . .

5 605606, 705606, 805606, . . .

6 75909, 75919, 75929, . . .

7 788620, 787620, 786620, . . .

8 64247, 64347, 64447, . . .

9 764940, 664940, 564940, . . .

10 512624, 522624, 532624, . . .

True or false?

11 One more than one hundred thousand, nine hundred and ninety-nine is two hundred thousand.

12 Sixty thousand is halfway between sixty-one thousand, four hundred and ninety-nine, and fifty-nine thousand, five hundred.

13 919191 < 919199

14 One million is one more than nine hundred and ninety-nine thousand, nine hundred and ninety-nine.

How many whole numbers are there between 1 million and 2 million?

I can count on and back from a number in steps of the same size

Negative numbers

Write the floor where each lift ends its journey.

1. −3

1 starts at 3, goes down 6 floors

2 starts at 1, goes down 3 floors

3 starts at 4, goes down 10 floors

4 starts at $^{-}8$, goes up 5 floors

5 starts at $^{-}2$, goes up 6 floors

6 starts at ground floor, goes down 1 floor

7 starts at 3, goes up 6, then down 4 floors

8 starts at 4, goes down 5, then up 3 floors

9 starts at 6, goes up 2, then down 9 floors

10 starts at 5, goes down 3, then up 4 floors

11 Write three more questions about the lift and also write the answers.

Dry ice is $^{-}35$°C in a freezer cabinet. If it warms up by $\frac{1}{2}$°C each day, how many days before it gets to 0°C? What if it rises $\frac{1}{4}$°C each day? $\frac{3}{4}$°C each day?

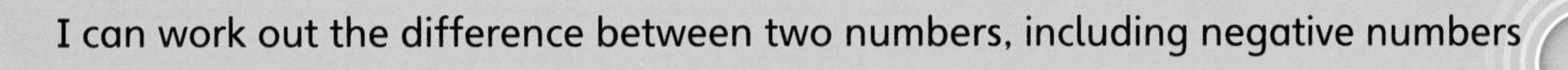

I can work out the difference between two numbers, including negative numbers

Negative numbers

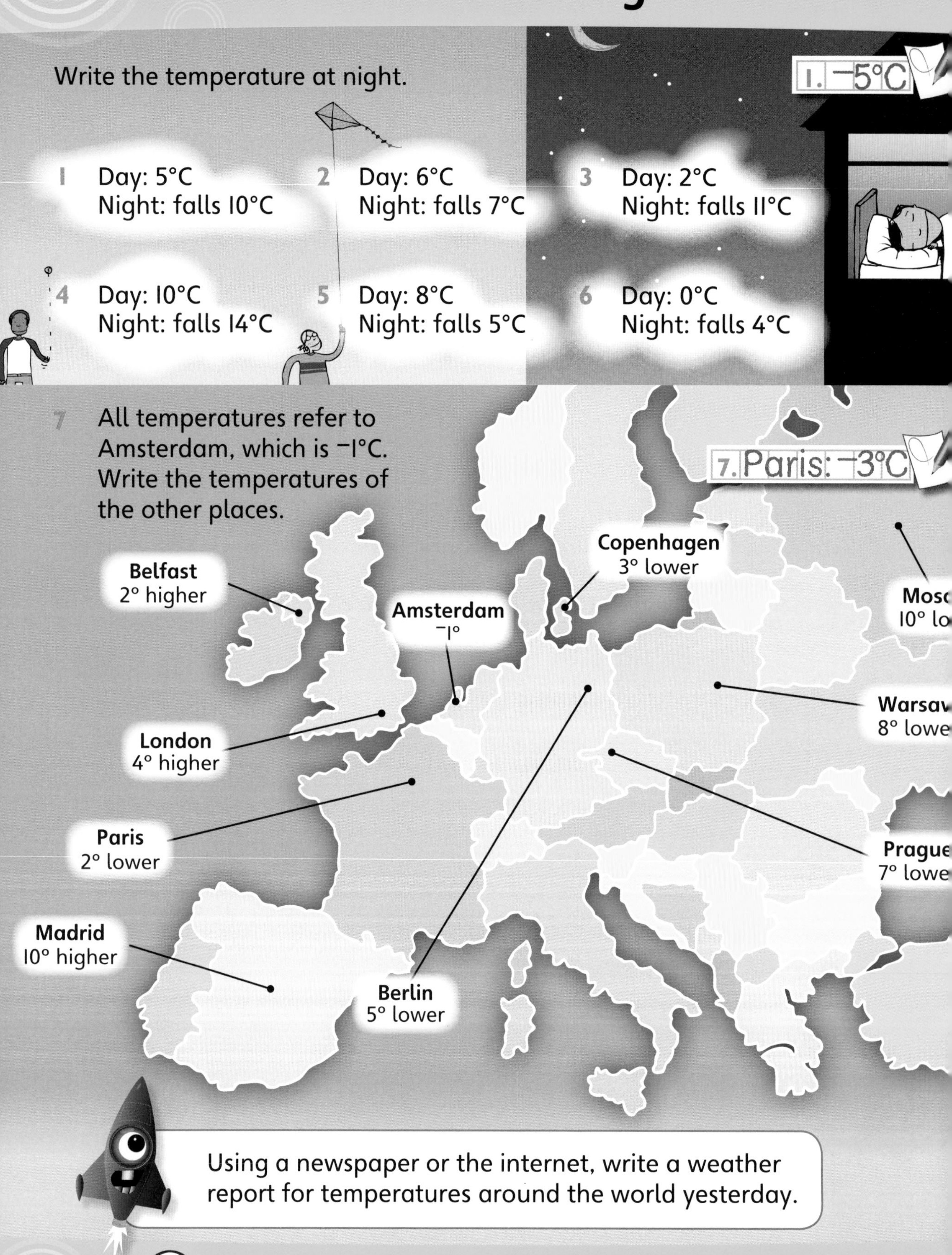

Write the temperature at night.

1 Day: 5°C
Night: falls 10°C

2 Day: 6°C
Night: falls 7°C

3 Day: 2°C
Night: falls 11°C

4 Day: 10°C
Night: falls 14°C

5 Day: 8°C
Night: falls 5°C

6 Day: 0°C
Night: falls 4°C

7 All temperatures refer to Amsterdam, which is −1°C. Write the temperatures of the other places.

Using a newspaper or the internet, write a weather report for temperatures around the world yesterday.

I can use negative numbers

Divisibility

The animals are going to the wildlife park.
Can they be paired exactly in 2s, yes or no?

1. No

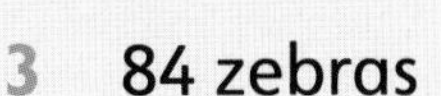

1 47 giraffes
2 56 horses
3 84 zebras
4 138 lions
5 249 tigers
6 54 elephants
7 427 llamas
8 109 monkeys
9 386 emus
10 245 kangaroos
11 1008 buffalo
12 478 snakes
13 97 rhinos
14 164 hippos
15 300 ostriches

For each set of animals that can be paired in 2s, write how many pairs. Can they be grouped exactly in 4s?

Write three numbers of animals that can be grouped in:

16. 6, . . .

16 2s, but not in 4s
17 5s, but not in 10s
18 5s, but not in 2s
19 4s, but not in 10s
20 Write three numbers of animals that cannot be grouped in 2s, 4s, 5s or 10s.

Investigate how many numbers between 1 and 100 are divisible by 2.
How many are divisible by 4? By 5?
How many are divisible by each of 2, 4 and 5?

I can use my halving skills to test the divisibility of a number

Tests of divisibility

1 Copy and complete the table. Write a tick to show that the number is divisible by the headings.

	2	3	4	5	6	8	9	10	25	50
140	✓			✓						
270										
3000										
85										
76										
432										
175										
234										
875										
4134										

2 Can you find any numbers that will have 6 ticks, 7 ticks, 8 ticks, 9 ticks, 10 ticks?

Matt thinks he has discovered a test for divisibility by 7 for 3-digit numbers, for example 637.

He says: double the hundreds digit. double 6 = 12

Then add it to the remaining part of the number. 12 + 37 = 49

If the resulting 2-digit number is divisible by 7, then so is the 3-digit number.

Investigate the method for different 3-digit numbers to test Matt's discovery.

I can use divisibility rules

Multiplying by 100 and 1000

Label the digits. Multiply by 100. Write the new value of the underlined digit.

1. HTU
361 × 100 = 36100
Value = 30 thousand

1	361	2	452	3	68	4	716
5	605	6	424	7	281	8	348
9	110	10	193	11	91	12	763

How many times must you multiply by 10 to get from 1 to 1 million?

Be a mathillionaire! Choose the correct answer.

13 430 × 10 is:
43 000 4303
430 4300

14 7602 × 100 is:
760 200 760 020
7 602 000 760 200

15 351 × 1000 is:
350 010 351 000
351 010 3 510 000

16 8080 × 1000 is:
8 080 000 8 080 100
808 000 8 080 080

17 5010 × 1000 is:
5 010 000 50 100
501 000 500 010

18 764 × 1000 is:
7640 764 000
76 400 76 060

I can multiply by 100 and 1000

Calculating with 10, 100 and 1000

1 Divide each red number by 1000. Find the blue number that matches your answer.

a) 288000	b) 360000	c) 3604	d) 288	e) 360400
f) 7110000	g) 637000	h) 63700	i) 490	j) 701
k) 3604000	l) 49000	m) 409	n) 49	o) 7110
p) 3606000	q) 40000	r) 90900	s) 637	t) 360
u) 9090000	v) 490000	w) 3606	x) 40	y) 9090

2 How many months are there in a century? In a millennium? How many weeks in a century? In a millennium?

3 A car costs 47p per mile to drive. How much does it cost to drive 100 miles? 1000 miles?

4 James saved £15 per week for 100 weeks. How much did he have? How much would he have after 2 years?

Copy and complete.

5. 273 × 100 = 27300

5 $273 \times 100 =$	6 $3600 \div 10 =$	7 $4875 \times 10 =$
8 $3140000 \div 1000 =$	9 $402 \times 1000 =$	10 $76 \times 1000 =$
11 $5138 \times 100 =$	12 $620 \times 10 =$	13 $53000 \div 100 =$

You multiply by 10 and your partner multiplies by 100. Start with a number such as 7. How many multiplications do you each do to get past 6 million?

I can multiply and divide by 10, 100 and 1000

2s, 3s, 4s, 5s, 10s

Write the next four multiples in each list.

1. 12, 14, 16, 18

1. 2, 4, 6, 8, 10
2. 3, 6, 9, 12, 15
3. 4, 8, 12, 16
4. 5, 10, 15, 20, 25
5. 10, 20, 30, 40

How many multiples of 2 under 50 are there?
How about multiples of 3? 4? 5?

6. Write down the multiples of 5 from this grid. Write the multiples of 10. Which numbers are in both lists?

31	32	33	34	35	36	37	38	39	40
41	42	43	44	45	46	47	48	49	50
51	52	53	54	55	56	57	58	59	60
61	62	63	64	65	66	67	68	69	70

7. Write down the multiples of 2 from this grid. Write the multiples of 3. Which numbers are in both lists?

1	2	3	4	5	6	7	8	9	10
11	12	13	14	15	16	17	18	19	20
21	22	23	24	25	26	27	28	29	30

I can list multiples of a number

Multiples

True or false?

1 All multiples of 4 are also multiples of 2.
2 All multiples of 5 are also multiples of 10.
3 All multiples of 2 are even numbers.
4 All multiples of 3 are odd numbers.
5 12 is a multiple of 2, 3, and 4.
6 All numbers which are multiples of 4 and 5 are multiples of 10.

1 point for a multiple of 2
2 points for a multiple of 5
3 points for a multiple of 3

7.	16	→	1 point
	21	→	3 points
	total		4 points

Write each overall score.

Write the smallest number that is:

13 a multiple of 2 and a multiple of 5
14 a multiple of 3 and a multiple of 2
15 a multiple of 5 and a multiple of 25
16 a multiple of 50 and a multiple of 100
17 a multiple of 4 and a multiple of 5
18 a multiple of 10 and a multiple of 25

I can find common multiples

Common multiples

Use the multiplication square to help you write all the numbers in the square that are multiples of:

1 7 2 4 3 9

1	2	3	4	5	6	7	8	9	10
2	4	6	8	10	12	14	16	18	20
3	6	9	12	15	18	21	24	27	30
4	8	12	16	20	24	28	32	36	40
5	10	15	20	25	30	35	40	45	50
6	12	18	24	30	36	42	48	54	60
7	14	21	28	35	42	49	56	63	70
8	16	24	32	40	48	56	64	72	80
9	18	27	36	45	54	63	72	81	90
10	20	30	40	50	60	70	80	90	100

Find the numbers that are common multiples of:

4 2 and 3
5 3 and 4
6 2 and 7
7 4 and 6
8 4 and 8
9 4 and 5

12 is a common multiple of 2, 3 and 4. Find another number that is a common multiple of consecutive numbers.

Find the smallest common multiples of:

10 2 and 3
11 2 and 5
12 3 and 4
13 3 and 5
14 2 and 4
15 3 and 12
16 4 and 6
17 8 and 12
18 6 and 8
19 8 and 10
20 10 and 15
21 25 and 30

I can compare multiples in different tables and find the common multiples

Factors

For each number, complete the pairs of factors.

1 20 → 1 × ☐, 2 × ☐, 4 × ☐

2 18 → 1 × ☐, 2 × ☐, 3 × ☐

3 30 → 1 × ☐, 2 × ☐, 3 × ☐, 5 × ☐

4 14 → 1 × ☐, 2 × ☐

5 40 → 1 × ☐, 2 × ☐, 4 × ☐, 5 × ☐

6 32 → 1 × ☐, 2 × ☐, 4 × ☐

7 54 → 1 × ☐, 2 × ☐, 3 × ☐, 6 × ☐

8 63 → 1 × ☐, 3 × ☐, 7 × ☐

Use your answers to write a list of factors of each number.

The numbers 30 and 40 each have four pairs of factors. Investigate other numbers with four pairs of factors.

Write a list of all the factors of these numbers.

9 6 10 16 11 10 12 50 13 28 14 48 15 60 16 52

Find a missing factor in each set.

17 Factors of 12: 1, 2, 6, 12, 4

18 Factors of 20: 1, 5, 20, 2, 10

19 Factors of 18: 1, 18, 6, 2, 3

20 Factors of 28: 1, 14, 4, 2

21 Factors of 15: 3, 5, 15

22 Factors of 40: 4, 1, 2, 8, 20, 10, 40

I can explain what a factor is and can work out the factor pairs of a number

Prime numbers

Large primes are used as security codes because they are difficult to crack.
Find some prime numbers greater than 100.
Try to find some very large ones.
Use the test for divisibility to help you.
Did you know? The largest prime number found so far has over 20 digits!

True or false?

1 All prime numbers are odd numbers.

2 There are ten prime numbers less than 30.

3 All prime numbers have exactly two factors.

4 The total of two prime numbers is always an even number.

5 Every number next to a multiple of 6 is a prime number.

6 Every 2-digit prime number is next to a multiple of 6.

7 Every 2-digit multiple of 6 is next to a prime number.

8 There is only one 2-digit prime number that has 6 as a tens digit.

9 1 is not a prime number.

10 All 2-digit prime numbers have a units digit of 1, 3, 7 or 9.

11 There are four prime numbers between 10 and 20.

12 A square number cannot be a prime number.

These pairs of prime numbers have a total of 90,
7 and 83
11 and 79
Can you find seven more pairs like this?

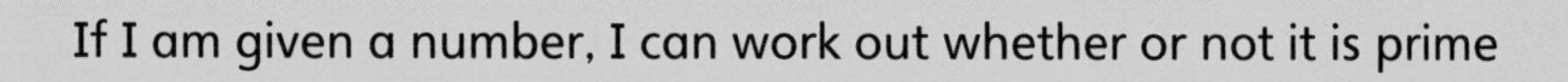

If I am given a number, I can work out whether or not it is prime

Re-ordering numbers

Re-order these numbers to make the calculations easier to do.

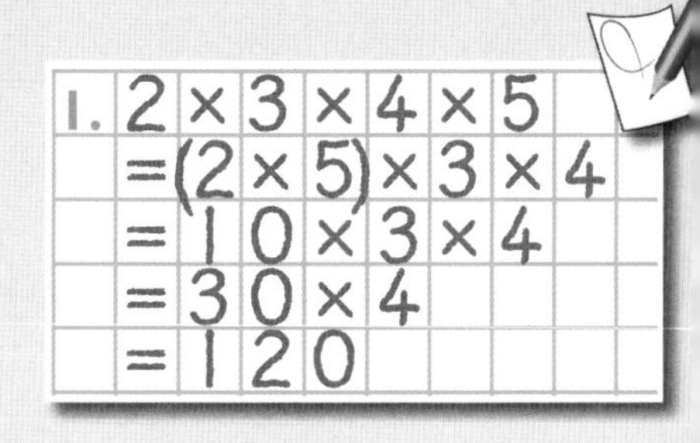

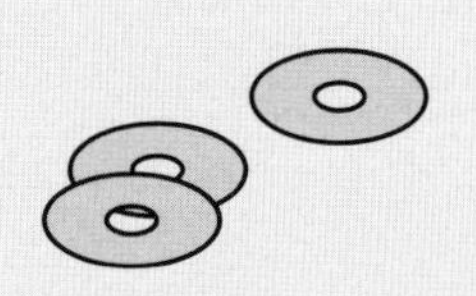

1. $2 \times 3 \times 4 \times 5 =$
2. $7 \times 2 \times 3 \times 5 =$
3. $4 \times 3 \times 5 \times 3 =$
4. $10 \times 2 \times 4 \times 3 =$
5. $9 \times 2 \times 4 =$
6. $5 \times 2 \times 5 \times 5 \times 5 \times 2 =$
7. $8 \times 3 \times 5 =$
8. $6 \times 7 \times 2 =$
9. $3 \times 5 \times 3 \times 6 =$
10. $3 \times 9 \times 6 \times 4 =$
11. $7 \times 3 \times 9 \times 4 =$
12. $4 \times 8 \times 7 \times 9 =$

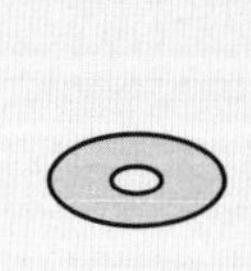

Make up some long calculations of your own to simplify.

I can re-order a set of numbers to make them easier to add

Using brackets

Each child has some vouchers. How much are they worth in total?

Hint: Choose your order. Use brackets to show which calculations you are doing together.

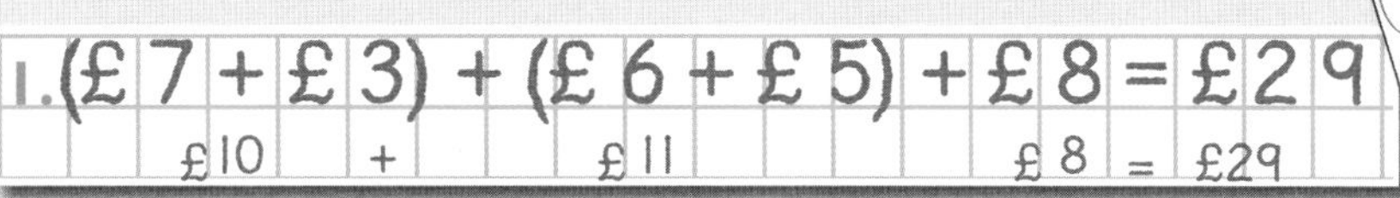

1. £5 £8 £7 £6 £3

2. £8 £8 £9 £6 £4

3. £5 £8 £3 £3 £6

4. £3 £5 £5 £6 £4

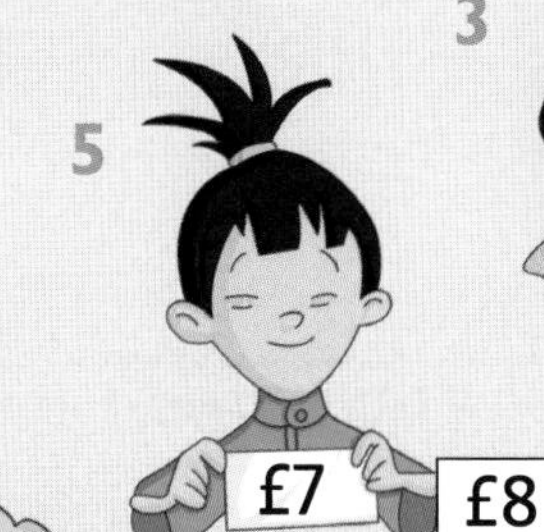

5. £7 £8 £9 £3 £4

6. £6 £8 £4 £5 £7

7. £8 £6 £6 £3 £5

8. £4 £8 £7 £4 £3

Rashida has five vouchers that have a total value of £40. What might each voucher be worth?

Complete these additions.

9.(8 + 2) + 9 + 6 + 7 = 32

Hint: Choose your order. Use brackets to show which calculations you are doing together.

9. 7 + 8 + 9 + 6 + 2 =
10. 8 + 9 + 7 + 5 + 9 =
11. 4 + 3 + 5 + 8 + 9 =
12. 7 + 6 + 3 + 4 + 8 =
13. 6 + 5 + 8 + 8 + 4 =
14. 5 + 3 + 6 + 6 + 4 =
15. 7 + 7 + 4 + 6 + 5 =
16. 3 + 5 + 5 + 7 + 8 =

I can use brackets to show how I have grouped together helpful numbers

Using brackets

Find each total.

1. (£40 + £60) + (£30 + £90) = £220
£100 + £120 = £220

Hint: Use brackets to show which calculations you are doing together.

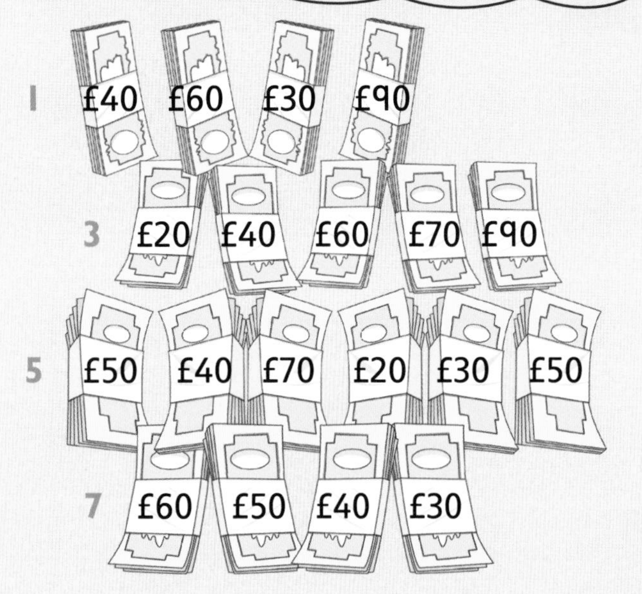

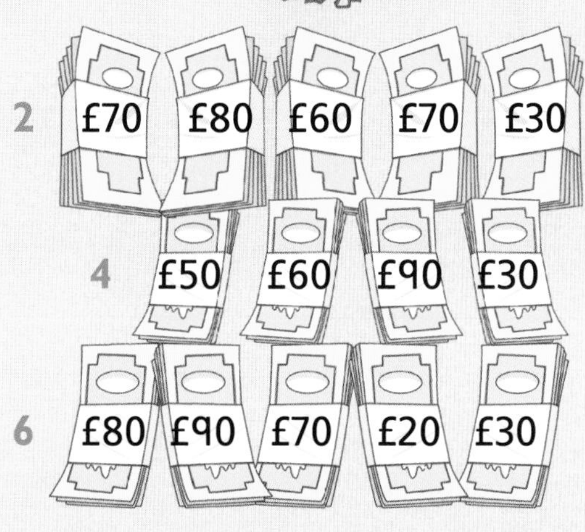

The table shows how much each plant grew each month. Use the table to answer the questions.

8 How much did each plant grow in the first 6 months of the year?

Plant	Jan	Feb	March	April	May	June
Bogsquash	60 cm	70 cm	30 cm	80 cm	20 cm	80 cm
Treacleflower	70 cm	90 cm	80 cm	60 cm	50 cm	30 cm
Googlygorse	40 cm	40 cm	60 cm	50 cm	70 cm	20 cm
Squiffleweed	30 cm	60 cm	70 cm	80 cm	50 cm	40 cm
Bumbleroot	80 cm	60 cm	50 cm	30 cm	70 cm	50 cm
Bogglerot	60 cm	50 cm	80 cm	60 cm	50 cm	40 cm

9 How much did all the plants grow in each month?

9. Bogsquash: 340 cm

If each plant grew 20 cm each month for the rest of the year, how much will each one have grown in a year?

I can use brackets to show how I group together helpful numbers

Using brackets

Write the total price. How much change do you get from £20?

Hint: Use the nearest multiple of 10. You can choose whether to count back or count on.

1. (£2 + £8) + (£5 + £4) = £19
£10 + £9 = £19
£1 change

1

2

3

4

In Nonsense Land there are only £3, £2 and £5 notes. Are there any amounts up to £20 that you cannot make?

Write the total amount.

5. 200 500 800 600

6. 500 700 800

7. 300 900 600 700

8. 400 300 800

9. 800 300 500 100

10. 600 400 300 500

I can use brackets to show how I group together helpful numbers

Using brackets and partitioning

Copy and complete.

1 $3 \times 43 = (3 \times 40) + (3 \times 3) = 120 + 9 = \_\_\_$

2 $4 \times 36 = (4 \times 30) + (4 \times 6) = \_\_\_ + \_\_\_ = \_\_\_$

3 $5 \times 27 = (5 \times 20) + (5 \times 7) = \_\_\_ + \_\_\_ = \_\_\_$

4 $9 \times 17 = (9 \times \_\_\_) + (9 \times \_\_\_) = \_\_\_ + \_\_\_ = \_\_\_$

5 $7 \times 52 = (7 \times \_\_\_) + (7 \times \_\_\_) = \_\_\_ + \_\_\_ = \_\_\_$

6 $6 \times 24 = (\_\_\_ \times \_\_\_) + (\_\_\_ \times \_\_\_) = \_\_\_ + \_\_\_ = \_\_\_$

7 $8 \times 32 = (\_\_\_ \times \_\_\_) + (\_\_\_ \times \_\_\_) = \_\_\_ + \_\_\_ = \_\_\_$

Write the cost of these jacket potatoes.

8 3 with beans

9 4 with sweetcorn

10 5 with mushy peas

11 7 with chilli

12 3 with sweetcorn and 4 with beans

8.
```
  32p
 × 3
  90     3 × 30
   6     3 × 2
  96p
```

Baked potato fillings

Beans 32p

Sweetcorn 28p

Mushy peas 42p

Chilli 46p

If four people order one jacket potato each, what could the total cost be?

I can use brackets to show how I group together helpful numbers

Using brackets and partitioning

Copy and complete.

1. $(6 \times 50) + (6 \times 6)$
$300 + 36 = 336$

1. $6 \times 56 =$
2. $8 \times 37 =$
3. $5 \times 78 =$
4. $7 \times 47 =$
5. $6 \times 64 =$
6. $7 \times 29 =$
7. $5 \times 23 =$
8. $3 \times 32 =$
9. $8 \times 34 =$

Write the total cost of the trip.

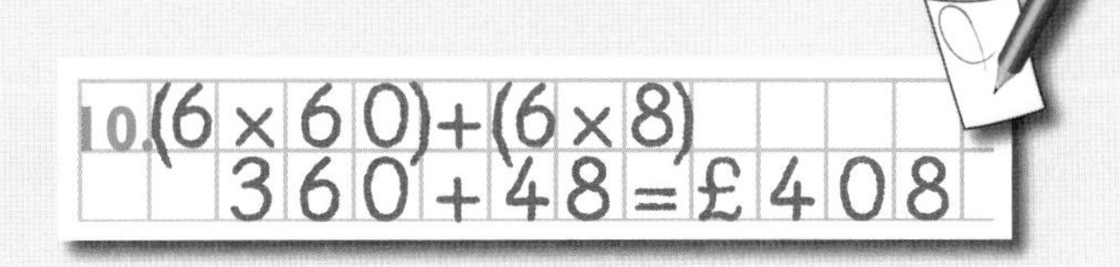

10 Fly to Paris £68

6 People

11 Fly to Rome £72

8 People

12 Sail to Jersey £44

7 People

13 Cycle to Dublin £36

7 People

14 Sail to Stranraer £56

8 People

15 Fly to Seville £59

6 People

16 Fly to Prague £22

6 People

17 Sail to Calais £87

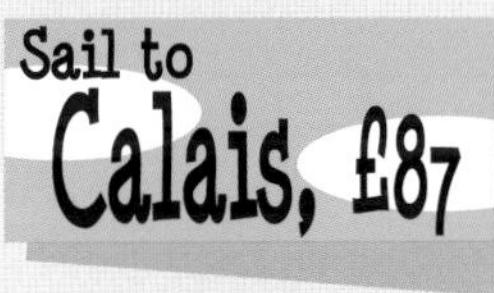

8 People

18 Fly to Milan £145

7 People

Work with a partner to create the 21 times-table.

I can show my thinking by using brackets when I solve a problem in several steps

Using brackets and partitioning

Complete the multiplications.

1 $3 \times 146 = (3 \times 100) + (3 \times 40) + (3 \times 6) = 300 + 120 + 18 =$

2 $5 \times 243 = (5 \times 200) + (5 \times 40) + (5 \times 3) =$

3 $4 \times 317 = (4 \times \square) + (4 \times \square) + (4 \times \square) =$

4 $6 \times 128 = (6 \times \square) + (6 \times \square) + (6 \times \square) =$

Complete the multiplications using brackets to help you.

5. $3 \times 416 = (3 \times 400) + (3 \times 10) + (3 \times 6)$
$= 1200 + 30 + 18$
$= 1248$

5 $416 \times 3 =$

6 $279 \times 5 =$

7 $186 \times 7 =$

8 $304 \times 6 =$

9 $512 \times 4 =$

10 $484 \times 3 =$

These are the sales for one Saturday at the Electronics Superstore. Write the amount collected.

11 £274

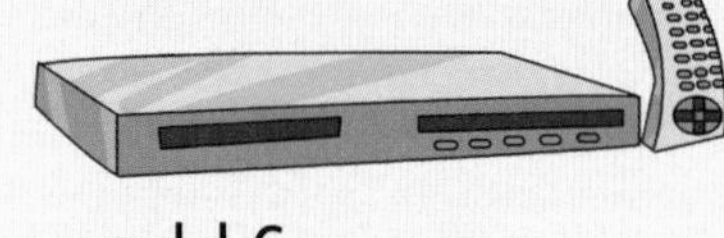

sold 6

12 £187

sold 3

13 £368

sold 4

14 £126

sold 8

15 £434

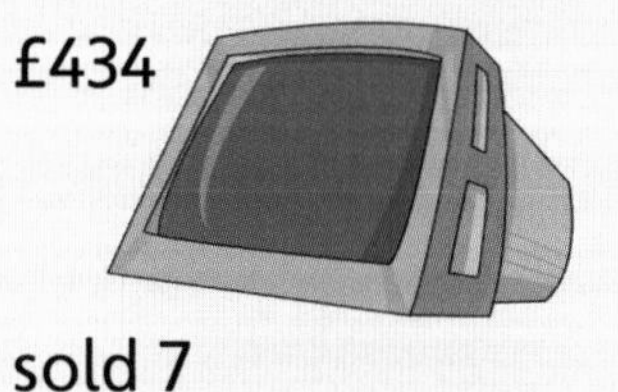

sold 7

16 £835

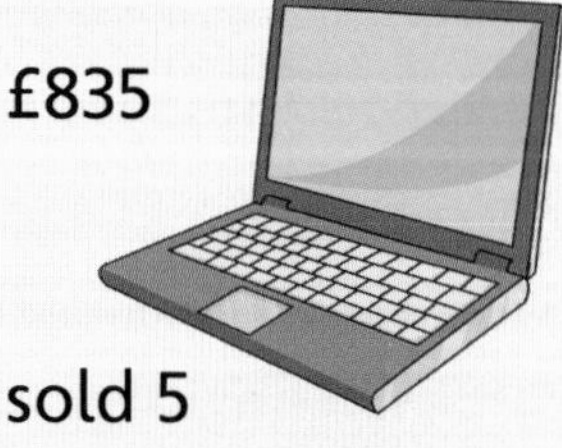

sold 5

How many of each item can you buy for £2000?

I can show my thinking by using brackets when I solve a problem in several steps

Using partitioning

Use the factors of the first number to help you do the multiplications.

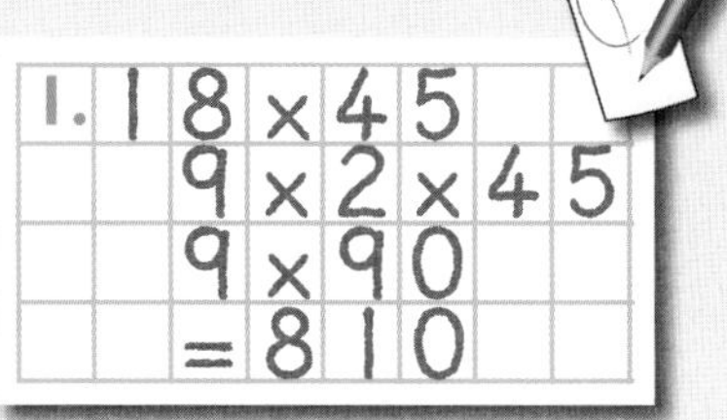

1 $18 \times 45 =$

2 $14 \times 35 =$

3 $14 \times 15 =$

4 $18 \times 55 =$

5 $16 \times 25 =$

6 $24 \times 35 =$

7 $15 \times 22 =$

8 $16 \times 42 =$

9 $16 \times 34 =$

Find the number of glasses in each box. What strategy will you use?

10

15 rows
22 glasses per row

11

18 rows
25 glasses per row

12

21 rows
19 glasses per row

13
16 rows
23 glasses per row

14
17 rows
25 glasses per row

15
18 rows
24 glasses per row

How could 480 glasses be arranged in a rectangular box?

I can show my thinking by writing out the steps when I solve a problem in stages

Using brackets and partitioning

Write out the 10 and the 2 times-tables.
Add the multiples.

Use your grid to help you work out the answers.

10	20	30	40	...
2	4	6	8	...
12	24	36	...	...

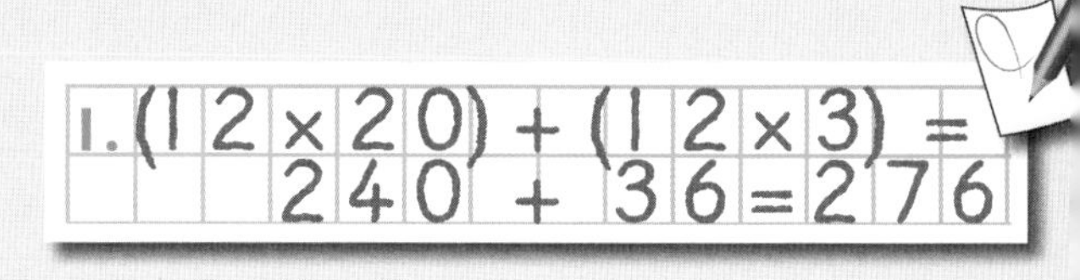

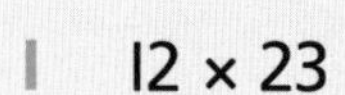
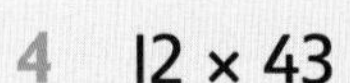

1. 12 × 23
2. 12 × 27
3. 12 × 31
4. 12 × 43
5. 35 × 12
6. 12 × 47
7. 52 × 12
8. 42 × 12
9. 39 × 12
10. 48 × 12
11. 33 × 12
12. 28 × 12

Work out the 13 times-table using the same method as shown above.
Use this table to help you multiply by 26.

What other multiplication facts can the 13 times-table help with? For example, × 39 or × 52.
Explore writing some really big multiplications, such as 8 × 52.

I can show my thinking by writing out the steps when I solve a problem in stages

Order of operations

Write these calculations with brackets.
Work out the answers and compare them.

1 $12 + 3 - 6 =$

2 $45 - 7 + 37 =$

3 $350 + 250 - 175 =$

4 $48 + 92 - 39 =$

5 $625 - 50 + 75 =$

6 $175 + 225 - 80 =$

Use the numbers 20, 16 and 38 with + and –.
What different maths questions can you make?

Use brackets to show the calculations you are carrying out first.
How many different calculations can you create by moving the brackets?

If you include one more number what possible combinations of calculations are there now?

Use the rule of operations to put these calculations into brackets and solve them.

7 $2 \times 10 + 4 \times 6 =$

8 $4 \div 2 + 6 \times 1 =$

9 $6 + 4 \times 3 + 1 =$

10 $7 - 3 \div 1 + 2 =$

11 $3 + 4 \div 2 + 5 =$

12 $60 + 40 \div 10 + 20 =$

I have investigated the rule of the order of operations

Rounding

Round each number to the nearest 100. Use the number line to help you.

1. 5983 → 6000

0 | 1000 | 2000 | 3000 | 4000 | 5000 | 6000 | 7000 | 8000 | 9000 | 10 000

1	5983	2	3812	3	3501	4	8712
5	6574	6	4329	7	2863	8	7750

Round each amount to the nearest £100.

17 Basanti's father said that whatever she saved by her birthday, he would give her the same amount rounded to the nearest £100. She has already saved £270. How much more must she save to get £400 from him?

18 If you did not have a number line explain how you could round to the nearest 1000, 100 and 10.

Write your own word problem involving rounding to the nearest 100.

I can round to the nearest 10 or 100

Rounding

Round the number of words to the nearest a: hundred and b: ten. You can draw a number line to help.

1 Oliver Twiddle
4328 words

2 Footy Facts
6795 words

3 Dinosaur Planet
3827 words

4 Classroom Mystery
13 452 words

5 Fun at the Stables
11 261 words

6 The School Legend
8875 words

7 Tiger Escape
6983 words

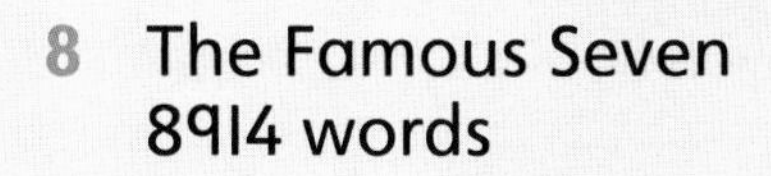

8 The Famous Seven
8914 words

9 Sleep Walker
9276 words

If Jayne reads these two books, find the total number of pages, then round it to the nearest hundred.

10 Oliver Twiddle and Fun at the Stables

11 Famous Seven and Dinosaur Planet

12 Sleep Walker and The School Legend

13 Footy Facts and Classroom Mystery

14 Tiger Escape and The School Legend

15 Oliver Twiddle and Sleep Walker

For each of questions 10–15, round both of the numbers to the nearest hundred, then add them.
Do you get the same answer as if you added first, then rounded?

I can round to the nearest 10 and 100 by looking at the digits

Rounding

Write the position of each pointer, then round the number to the nearest:

1.a: 6560 → 7000

1 — number line from 6000 to 7000, pointers d, b, a, e, c — thousand

2 — number line from 5300 to 5400, pointers f, i, g, h, j — hundred

3 — number line from 2700 to 2800, pointers m, p, k, n, l — ten

Round each price to the nearest a: thousand pounds b: hundred pounds and c: ten pounds.

4. a: £8146 → £8000
b: £8100
c: ...

4 £8146

5 £7234

6 £3974

7 £9148

8 £5158

9 £11762

10 £12349

11 £6695

12 £15685

Write a 4-digit number which gives the same rounded number when rounded to the nearest thousand, hundred and ten.

I can round to the nearest 10, 100 or 1000

Rounding

These are the attendances at some football matches.
Round them to the nearest a: thousand and b: hundred.

1 Rovers
27 564

2 City
18 546

3 Albion
43 582

4 United
13 712

5 Athletic
64 789

6 Wanderers
34 358

Find last Saturday's attendance figures at a football game.
Round them to the nearest thousand and the nearest hundred.

7 At last Saturday's match there were 7652 adults and 2847 children. Find the total number of spectators. Round your answer to the nearest hundred. If each person bought a programme for £3, how much money was made?

8 

A club sold their striker for a fee of £51 465 and paid £37 374 for a new goalkeeper. To the nearest thousand, approximately how much money did the club gain?

Write the a: smallest and b: largest possible number before it was rounded.

9 3700 rounded to the nearest hundred

10 75 000 rounded to the nearest thousand

11 4860 rounded to the nearest ten

12 469 000 rounded to the nearest thousand

13 58 700 rounded to the nearest hundred

14 47 390 rounded to the nearest ten

I can round to the nearest 10, 100 or 1000

Working backwards and forwards

Fill in the missing numbers.

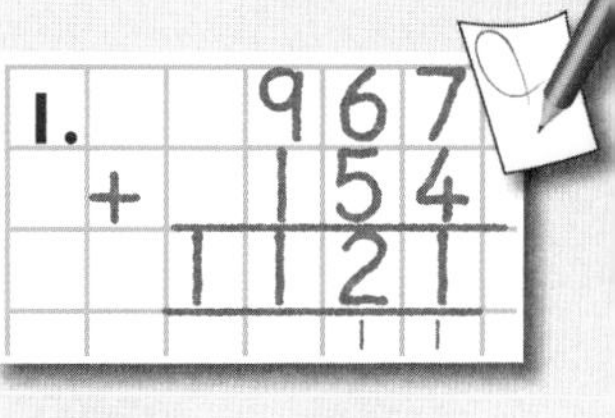

1

   9 6 7
+ 1 5 □
□ □ 2 1
  1 1

2

  □ 2 4 6
+ 9 2 □ 3
1 3 □ 2 □
  1

3

  5 7 6 9
+ □ □ 8 4
1 3 5 □ 3
  1 1 1

Fill in the missing numbers.

4

  4 7 5 8
+ □ 5 8 6
1 2 □ □ □

5

  6 3 7 2
+ 7 □ 9 6
□ □ 2 □ □
  1 1

6

  8 7 6 5
+ 9 4 □ □
□ □ □ 4 0
  1 1 1

7

  7 8 5 3
+ 3 7 □ 6
1 □ □ 9 9
  1

8

  6 4 8 2
+ □ □ 9 6
1 2 2 □ □
  1 1

9

  5 4 9 7
+ 8 6 □ 5
□ □ 1 2 □
  1 1 1

10 Now make up 3 more calculations for a friend to solve.

Each letter is a single digit. No digit is represented by more than 1 letter, for example, if N = 2 no other letter can be 2. Give each letter a digit to make this addition work:

  R A I N
+ S N O W
S L E E T

I can check answers to addition and subtraction calculations by using the inverse operation

Adding and subtracting

Find the difference between the number of miles travelled by each car.

Hint: Use the nearest multiple of 10. You can choose whether to count back or count on.

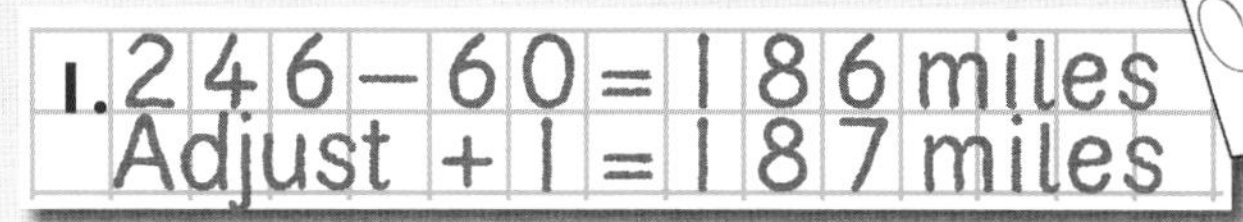

1

246 miles 59 miles

2

427 miles 79 miles

3

154 miles 88 miles

4

285 miles 79 miles

5

274 miles 57 miles

6

164 miles 89 miles

How many centimetres must each snail crawl to reach the lettuce?

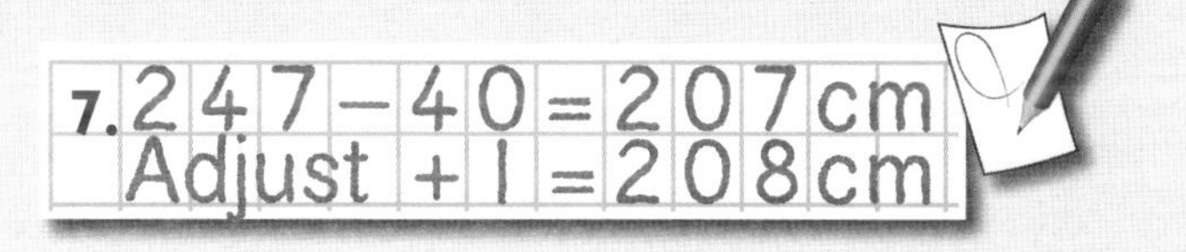

7 **8** **9** **10** **11**

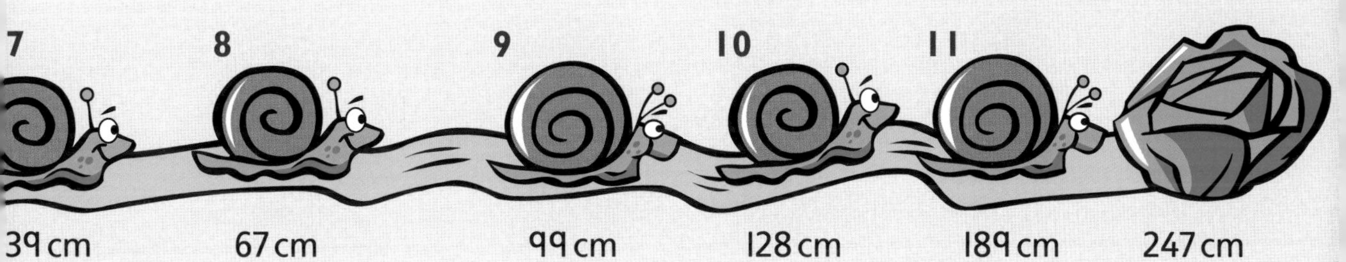

39 cm 67 cm 99 cm 128 cm 189 cm 247 cm

Work with a partner. Agree who will count back and who will count on. Try this subtraction: 167 – 89. Reverse roles. Which is the best method for you? Try 224 – 129.

I can do a calculation by rounding one of the numbers and then adjust my answer at the end

Counting on and back to calculate

1 Choose three cards to make six different totals close to 900.

2 Choose four cards to make three different totals between 1000 and 1200.

True or false?

3 When adding five numbers, if the units digits are all the same, the total ends in 5.

4 Three different numbers are added. The total is over 1800. All three numbers must be 3-digit numbers.

5 Adding four 3-digit numbers less than 500 cannot give a total over 2000.

6 It is possible to add another number to the sum of 373 and 737 so that the total has four identical digits.

7 The total of four 2-digit numbers, where all of the digits are odd, must be an odd number.

Add two palindromic 3-digit numbers, such as 242 and 353. Is the answer a palindrome? Try some more.

I can count on to the next 10, 100 or 1000 and can use this to solve calculations

Adding

1 Find pairs of flags that make 100.
Write the addition. Do this 10 times.

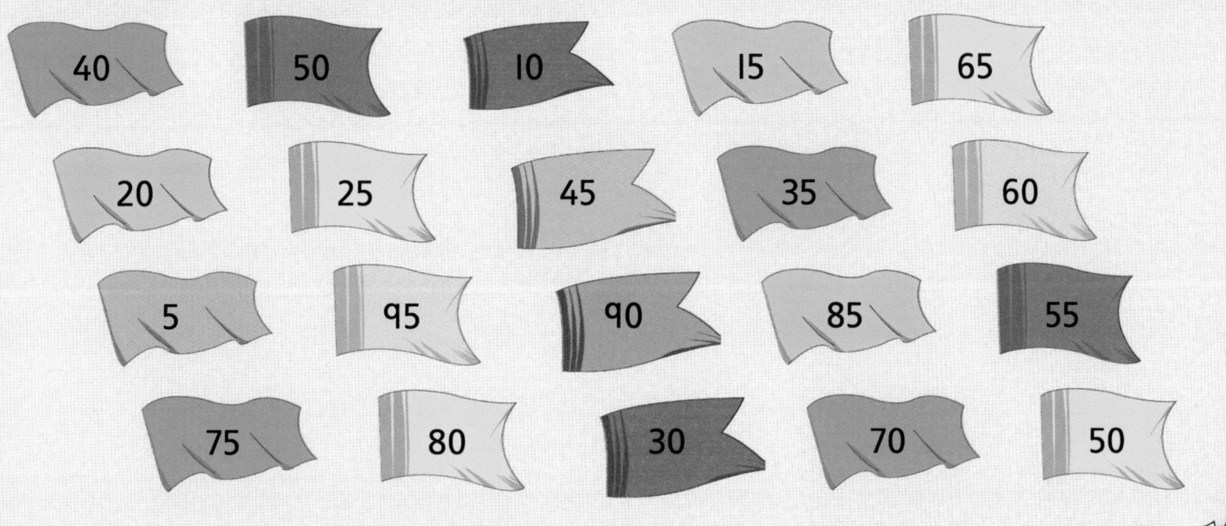

How long until each person reaches the next multiple of 10 years?

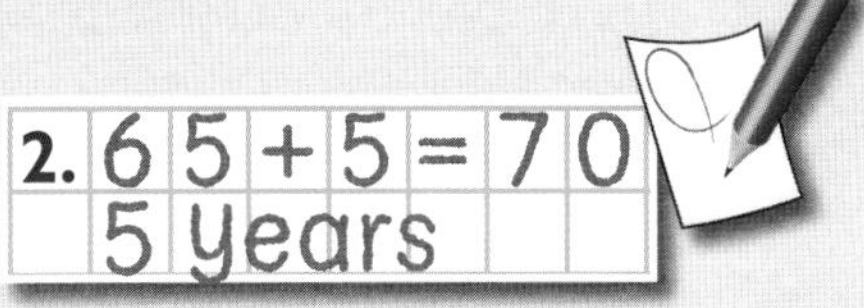

2

65 years old

3

18 years old

4

27 years old

5

43 years old

6

32 years old

7

87 years old

8

51 years old

9

44 years old

How many ways can you find of making £1 using 10p, 20p and 50p coins?

Adding and subtracting

Each athlete has run part of a 1000 m race. How far is there left to run?

1. 350 + 650 = 1000 m

1	350 m	**2**	450 m	**3**	650 m
4	850 m	**5**	750 m	**6**	420 m
7	680 m	**8**	730 m	**9**	230 m

Three multiples of 50 total 1000. What could they be? Can you find all the possibilities?

10 Taife and Joe share 1000 g of dog food. Taife has a big dog so she takes 660 g. How much does Joe have?

11 Ashley squeezes oranges into a litre jug. One litre is a thousand millilitres. He uses 10 small oranges to get 780 ml. How much more juice will he need to fill the jug?

12 Tanvi has saved £68. How much more must she save to have £100? Her sister has saved £54. How much more does she need?

Work out the number of pounds that must be added to each of these to make £1000: £362, £458, £671, £884. Can you find a quick way of doing this?

I can work out what to add on to make a multiple of 10 or 100

Doubling and halving

Double these numbers by doubling the tens, doubling the units, then combining.

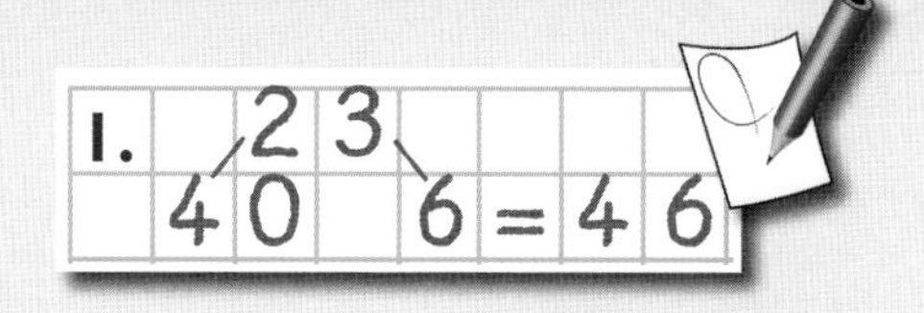

1 23 2 31 3 44 4 12

5 16 6 27 7 38 8 19

9 28 10 36 11 49 12 57

Halve these numbers by halving the tens, halving the units, then combining.

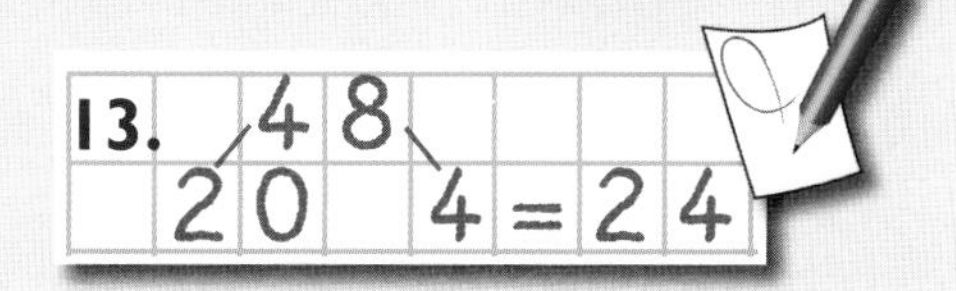

13 48 14 26 15 46 16 82

17 34 18 58 19 76 20 38

21 92 22 64 23 72 24 54

Write the cost of two of each item.

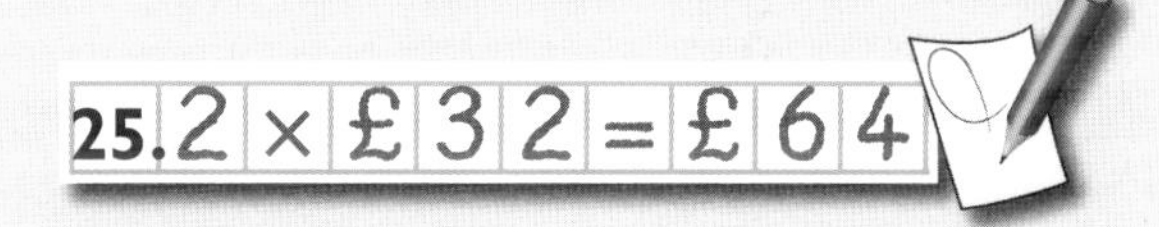

25 

26 

27 

28 

29 

Write the cost of each item in a half-price sale.

Find the cost of five of each item by halving the cost of ten of each. How could you find the cost of 20 of one of the items?

I can separate a number into parts in order to halve or double it

Near doubles

Use doubling to complete these additions.

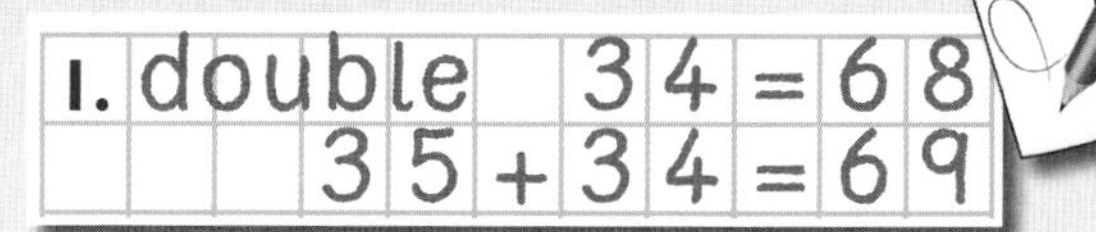

1 double 34 =
34 + 35 =

2 double 42 =
42 + 41 =

3 double 26 =
26 + 27 =

4 double 18 =
18 + 19 =

5 double 23 =
23 + 22 =

6 double 45 =
45 + 44 =

7 double 28
28 + 27 =

8 double 37 =
37 + 36 =

9 double 46 =
46 + 47 =

Write a near double for each of these numbers.

10. double 31 + 1

10 63 11 31 12 45 13 29

14 87 15 53 16 71 17 95

Guess who I am:

18 I am a number less than 20. When I am doubled the answer is 6 multiplied by itself.

19 I am a number greater than 40. When I am halved the answer is double 12.

20 I am a number less than 15. When I am doubled the answer is the seventh multiple of 4.

21 I am a number greater than 50. When I am halved the answer is the number of days in March.

Invent some 'Guess who I am' problems.

I can use 'near doubles' to add two next-door numbers

Using doubles and halves

There are 50 paperclips in a box.
How many paperclips in:

1. 13 × 100 = 1300
   13 × 50 = 650

1 13 boxes

2 22 boxes

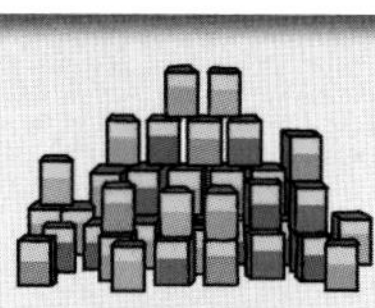

3 38 boxes

4 17 boxes

5 41 boxes

6 27 boxes

7 35 boxes

8 19 boxes

9 24 boxes?

There are 25 candles in a box.
How many candles in:

10. 16 × 100 = 1600
    16 × 50 = 800
    16 × 25 = 400

10 16 boxes

11 34 boxes

12 28 boxes

13 22 boxes

14 36 boxes

15 44 boxes

16 58 boxes

17 64 boxes

18 72 boxes?

Invent a method for multiplying by 150 using doubling and halving. Write some multiplications by 150 for your partner to answer.

I can use my halving skills to multiply by 50 and 25

Adding

Jamie

Kulpreet

Ling

Emma

Mark

Find the total scores for:

1.	8 3 6
+	3 4 2
	1 1 7 8

1 Jamie and Kulpreet

2 Ling and Emma

3 Emma and Mark

4 Ling and Kulpreet

5 Out of all the children, which pair has the largest score? What is it?

6 Out of all the children, which pair has the lowest score? What is it?

Find the totals.

7.	4 6 5 0
+	3 7 2 5
	8 3 7 5
	1

7 $\begin{array}{r} 4650 \\ +\ 3725 \\ \hline \end{array}$

8 $\begin{array}{r} 5671 \\ +\ 3157 \\ \hline \end{array}$

9 $\begin{array}{r} 3712 \\ +\ 2834 \\ \hline \end{array}$

10 6172 + 2519 =

11 4384 + 3575 =

12 3942 + 1436 =

13 6742 + 2637 =

14 4235 + 3417 =

What digits can make this calculation work?

$$\begin{array}{r} \square\ 3\ \square\ 3 \\ +\ 3\ \square\ 3\ \square \\ \hline 1\ 0\ 0\ 0\ 0 \\ \hline \end{array}$$

I can explain how I do a written addition calculation

Adding

Find the total number of people at each match. Write an estimate first.

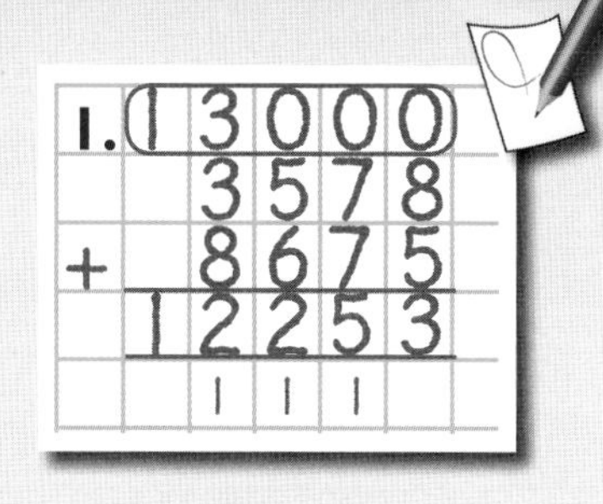

1 Three thousand, five hundred and seventy-eight children and eight thousand, six hundred and seventy-five adults.

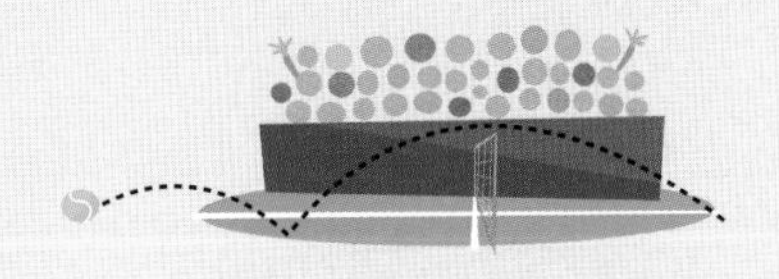

2 Six thousand, nine-hundred and ninety-five children and two hundred and eighteen adults.

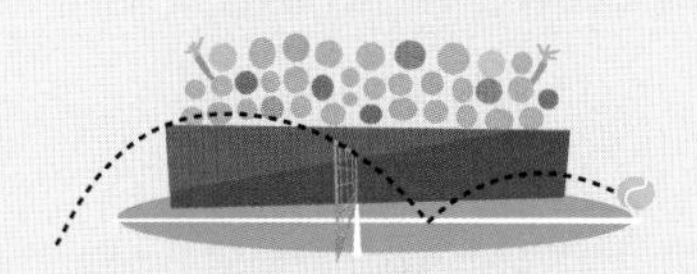

3 Three thousand, four hundred and forty-three children and six thousand, three hundred and fifty-four adults.

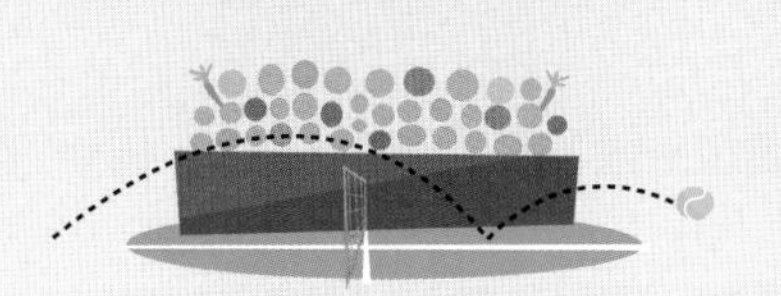

4 Seven thousand, six hundred and sixty-eight children and eight thousand, nine hundred and eighty-five adults.

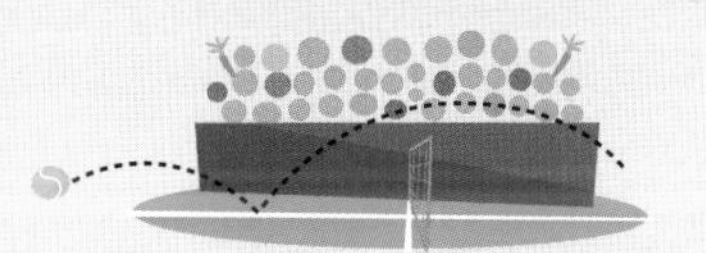

Arrange the cards 1–8 to make an addition of two 4-digit numbers. What is the largest odd total you can make? What is the smallest? What is the largest even total you can make? What is the smallest?

Copy and complete, then write your answer in words.

5 4783 + 5878 = 6 6786 + 8595 = 7 3642 + 6272 =

8 8872 + 3569 = 9 7535 + 8169 = 10 1368 + 9453 =

I can explain how I do a written addition calculation

Subtracting

How much further to go?

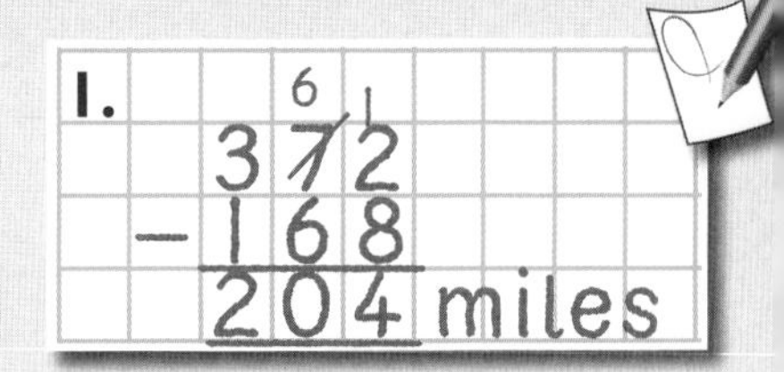

1 

Journey 372 miles
Done 168 miles

2 

Journey 683 miles
Done 329 miles

3 

Journey 492 miles
Done 227 miles

4 Journey 871 miles
Done 436 miles

5 Journey 683 miles
Done 325 miles

6 Journey 792 miles
Done 244 miles

7 Journey 983 miles
Done 826 miles

8 Journey 674 miles
Done 238 miles

9 Journey 462 miles
Done 219 miles

How far has the hot air balloon fallen?

10 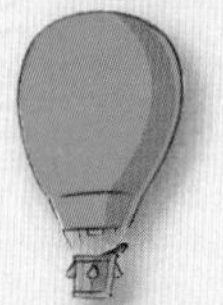

Was 816 m
Now 542 m

11 

Was 907 m
Now 261 m

12 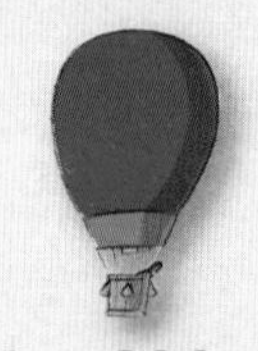

Was 828 m
Now 453 m

13 

Was 719 m
Now 256 m

14 

Was 836 m
Now 382 m

15 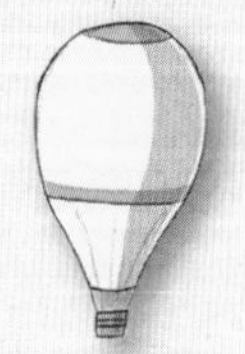

Was 948 m
Now 295 m

16 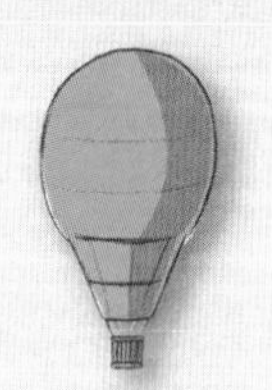

Was 827 m
Now 586 m

17 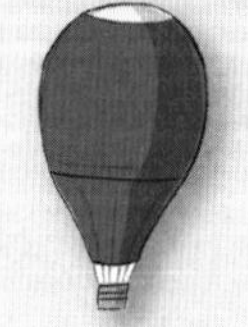

Was 639 m
Now 265 m

What is the largest 3-digit number that can be added to two other 3-digit numbers to make 999?

I can explain how I solve a subtraction problem using formal methods

Adding

Complete these additions.

1

```
   4683
    742
   3604
 +   28
 ------
```

2

```
   3568
     47
 +  362
 ------
```

```
1.   4683
      742
     3604
 +     28
     ----
     9057
     211
```

3 1462 + 556 + 98 + 1134 =

4 2673 + 843 + 62 + 359 =

5 6437 + 362 + 12 + 1794 =

6 38 + 691 + 3742 + 6438 =

7 Choose four trips. Add the prices to find the total cost. Repeat 10 times.

Fly to Paris
£87

Weekend in
Monte Carlo
£368

Las Vegas
£2565

Sun & Surf
Down Under
£5677

Wildlife Galore
in South Africa
£1858

Icelandic cruise
£4667

Tour
America
£8874

5-day break in
Amsterdam
£475

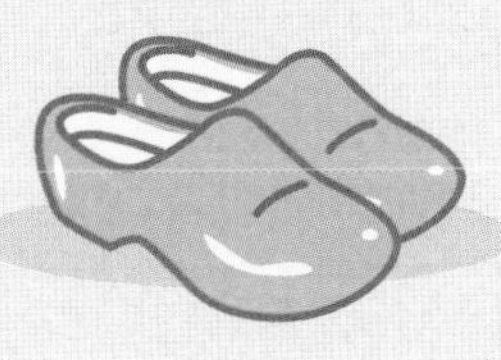

You have £5000. Which of these trips could you choose?

Subtracting

Subtract the smaller from the larger number.

1 5432 – 2118 =

2 4826 – 1452 =

3 6128 – 4029 =

4 7269 – 3542 =

5 7636 – 2172 =

6 6583 – 4128 =

7 5296 – 2734 =

8 4831 – 3614 =

Work out the differences between the two distances. Then write your answer in words.

9 3175 km 1432 km

10 5647 km 2518 km

11 4361 km 9628 km

12 8496 km 3753 km

13 3271 km 7632 km

14 2219 km 9438 km

15 5296 km 2534 km

16 5272 km 8635 km

Arrange the digits 1–8 in this calculation:

☐☐☐☐ – ☐☐☐☐ =

to give the largest possible answer.

I can explain how I solve subtraction problems using formal methods

Subtracting

Write the difference between the two amounts.

1

2

3

4

5

6

7

8

9

10

11

12

Work with a partner. Choose two of the subtractions each. Check your partner's answers by adding the answer to the smaller number.

Copy and complete.

13 53146 − 2817 = ____

14 3881 − 1976 = ____

15 64378 − 2769 = ____

16 5436 − 1787 = ____

17 33265 − 1678 = ____

18 6384 − 2637 = ____

Subtracting

Write the difference between the mountain heights.

Mont Blanc	Eiger	Matterhorn	Weisshorn	Silberhorn
4810 m	3974 m	4478 m	4505 m	3695 m

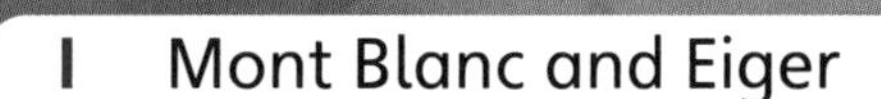

1 Mont Blanc and Eiger

2 Matterhorn and Weisshorn

3 Silberhorn and Matterhorn

4 Weisshorn and Mont Blanc

5 Eiger and Matterhorn

6 Mont Blanc and Silberhorn

7 Weisshorn and Silberhorn

8 Eiger and Weisshorn

9 Mont Blanc and Matterhorn

10 Silberhorn and Eiger

Use an atlas to work out the differences in height between Mount Everest and some of the other mountains in the Himalayas.

Copy and complete. Then check your answers with a calculator.

11 46831 – 7994 =

12 5218 – 3961 =

13 7654 – 379 =

14 64382 – 5317 =

15 47932 – 8364 =

16 6431 – 3786 =

17 32465 – 6374 =

18 4261 – 1794 =

19 47653 – 24278 =

20 22643 – 1875 =

21 34652 – 26741 =

22 18567 – 9379 =

I can explain how I solve difference problems

Multiplying

Complete these multiplications.

Hint: Multiply the nearest multiple of 10 first. Then choose whether to count on or count back to find your answer.

1. 20 3
5
← 23 →
(5 × 20) + (5 × 3)
100 + 15
= 115

1 23 × 5 =
2 52 × 7 =
3 99 × 4 =
4 101 × 6 =
5 51 × 3 =
6 49 × 8 =

7 Jason scores 23 points in each match and he plays 9 matches in a season. How many points does he score?

8 Parvati has 13 shots on target each game and she plays 8 games How many shots on target does she have in all?

Oranges are packed in boxes of 49. Write the total number of oranges.

9 
4 boxes

10  
6 boxes

11 
7 boxes

12 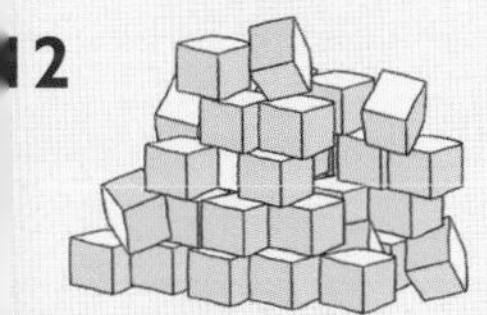
2 boxes

13 
9 boxes

14 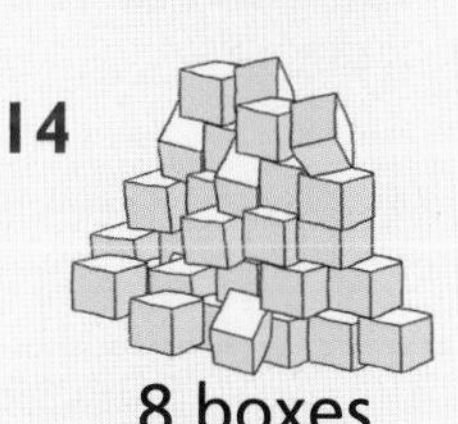
8 boxes

15 
3 boxes

16 
5 boxes

17
11 boxes

I can use rounding to help me complete calculations

Multiplying by multiples of 10

Children are having a sponsored remote control car challenge around their gardens. How far has each car gone after 40 laps?

Which method will you choose?

Grid:

	10	10	10
10	100	100	100
10	100	100	100
10	100	100	100
10	100	100	100
		1200	

or partition:

$$4 \times 30 = 4 \times 10 \times 3 \times 10$$
$$= 4 \times 3 \times 10 \times 10$$
$$= 12 \times 100$$
$$= 1200$$

1 

1 lap = 30 m

2 

1 lap = 40 m

3 

1 lap = 60 m

4 

1 lap = 70 m

5 

1 lap = 20 m

6 

1 lap = 90 m

Children put long-life batteries in their cars. How far has each one travelled after 80 laps? What do you notice about your answers?

Copy and complete.

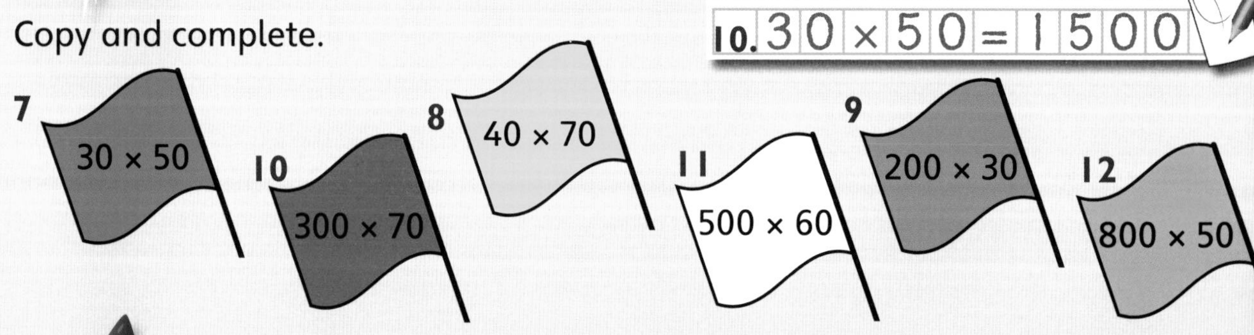

Your friend does not know how to solve 800 × 50. How would you teach them to do this?

I can use my knowledge of table facts and multiplying by ten to work out new facts

Multiplying

Copy and complete these multiplication grids. Write a multiplication for each one.

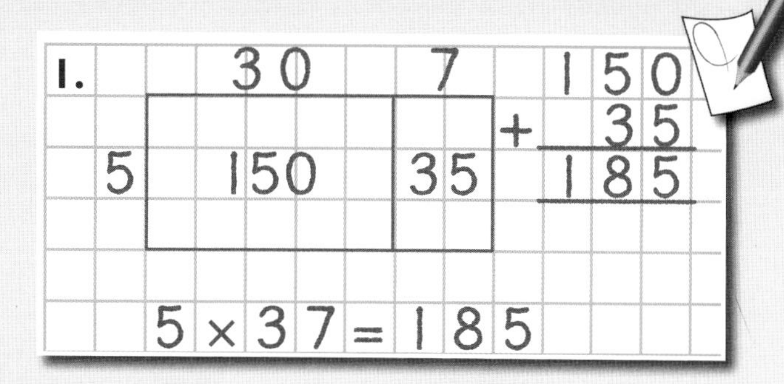

1

30 7

5

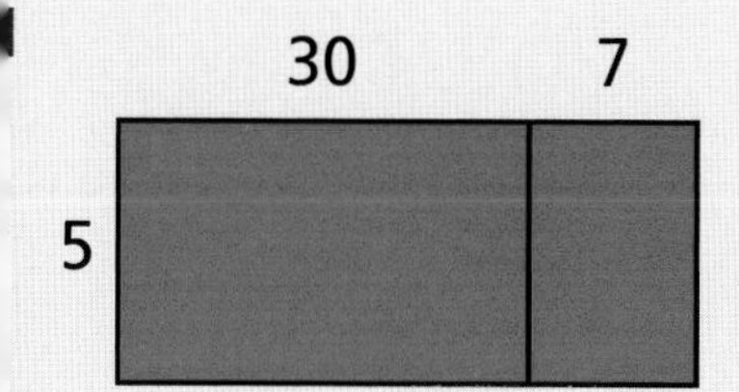

2

40 3

6

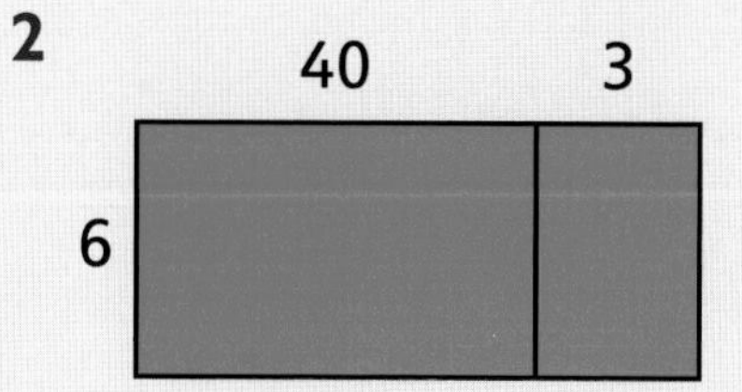

3

20 8

3

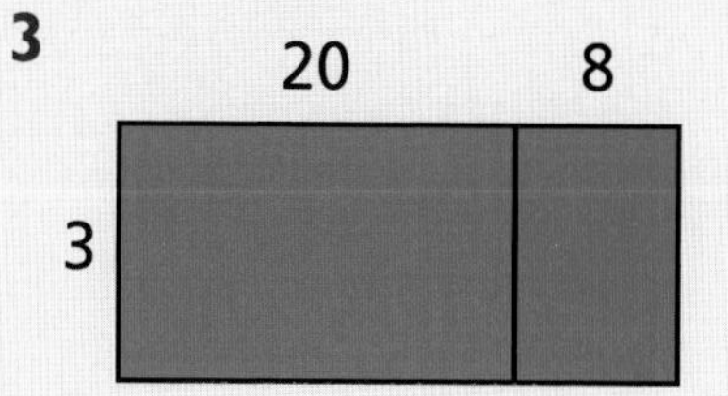

4

70 2

4

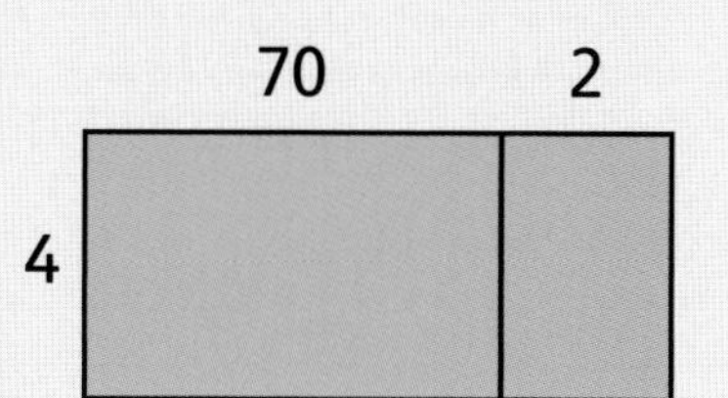

5

30 4

8

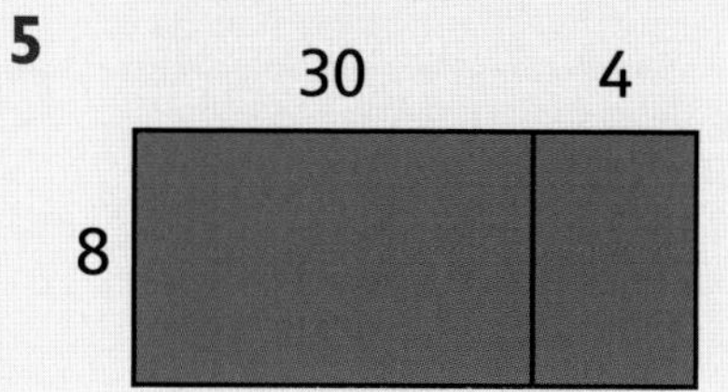

6

40 2

9

Complete these multiplications. Estimate first, then draw a grid.

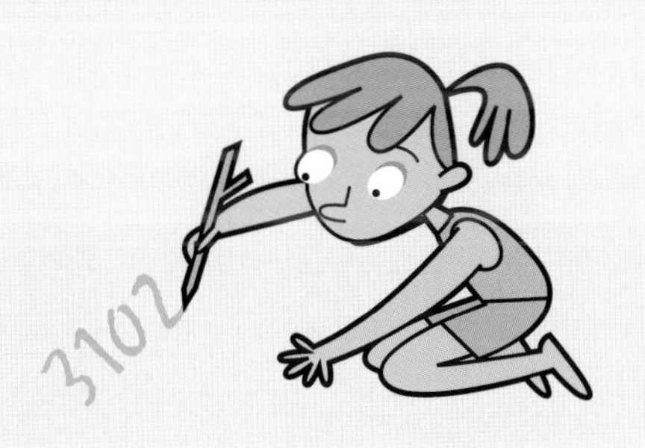

7. estimate: 3 × 30 = 90

20 7

3 60 21

3 × 27 = 81

7 3 × 27 =

8 4 × 43 =

9 5 × 38 =

10 6 × 74 =

11 7 × 33 =

12 8 × 29 =

Can you find a multiplication like this that has an answer near to 236?

I can use the grid method to multiply a 2-digit number by a 1-digit number

Multiplying

1 Copy and complete this multiplication table.

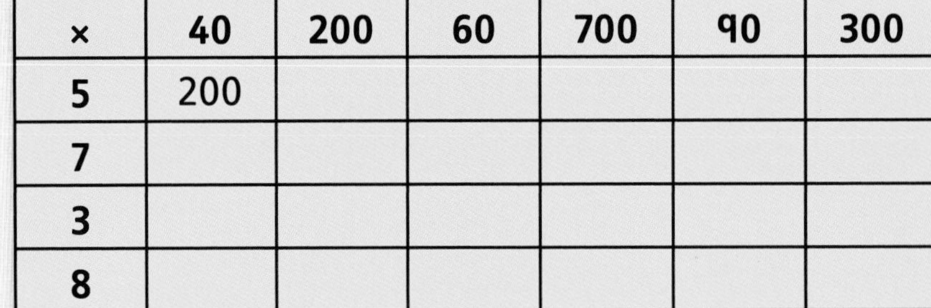

×	40	200	60	700	90	300
5	200					
7						
3						
8						

Investigate what multiplications will have an answer of 2400.

Copy and complete.

2.	200	70	5		600
3	600	210	15		210
				+	15
					825

275 × 3 = 825

2 200 70 5

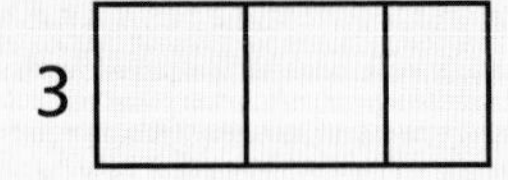

275 × 3 = ★

3 300 40 2

342 × 4 = ★

4 400 20 3

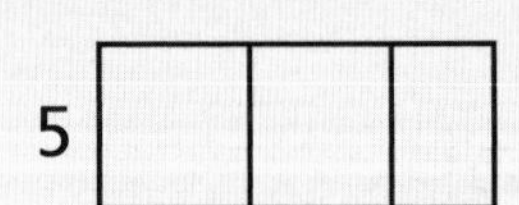

423 × 5 = ★

5 158 × 6 = ★

6 439 × 5 = ★

7 618 × 2 = ★

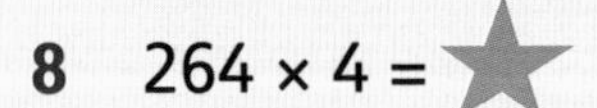

8 264 × 4 = ★

9 397 × 4 = ★

10 561 × 3 = ★

11 482 × 6 = ★

12 279 × 5 = ★

13 845 × 2 = ★

14 659 × 4 = ★

15 354 × 3 = ★

16 415 × 6 = ★

I can use the grid method to multiply a 3-digit number by a 1-digit number

Multiplying

Are these answers correct? Check using the grid method.
Write the correct answer.

1 348 × 6 = 2188

2 562 × 3 = 1686

3 238 × 7 = 1466

4 618 × 5 = 3095

5 279 × 7 = 1956

6 328 × 4 = 1312

7 597 × 4 = 3628

8 592 × 3 = 1786

9 283 × 6 = 1698

James and Padma are talking about their maths. James did 340 x 6 and got the answer 1824. Padma got 2040. Who is correct and what mistake was made?

Now try these.

10 209 × 8 =

11 380 × 7 =

12 406 × 7 =

13 705 × 8 =

14 470 × 3 =

15 608 × 4 =

16 504 × 9 =

17 307 × 4 =

18 790 × 8 =

19 Paving slabs are 108 cm wide. Nine slabs are laid in a row. How wide will this be? The slabs are meant to fill a space 10 m wide. Ten metres is the same as 1000 centimetres. How much space is left to fill?

I can lay out multiplication calculations in a grid

Multiplying

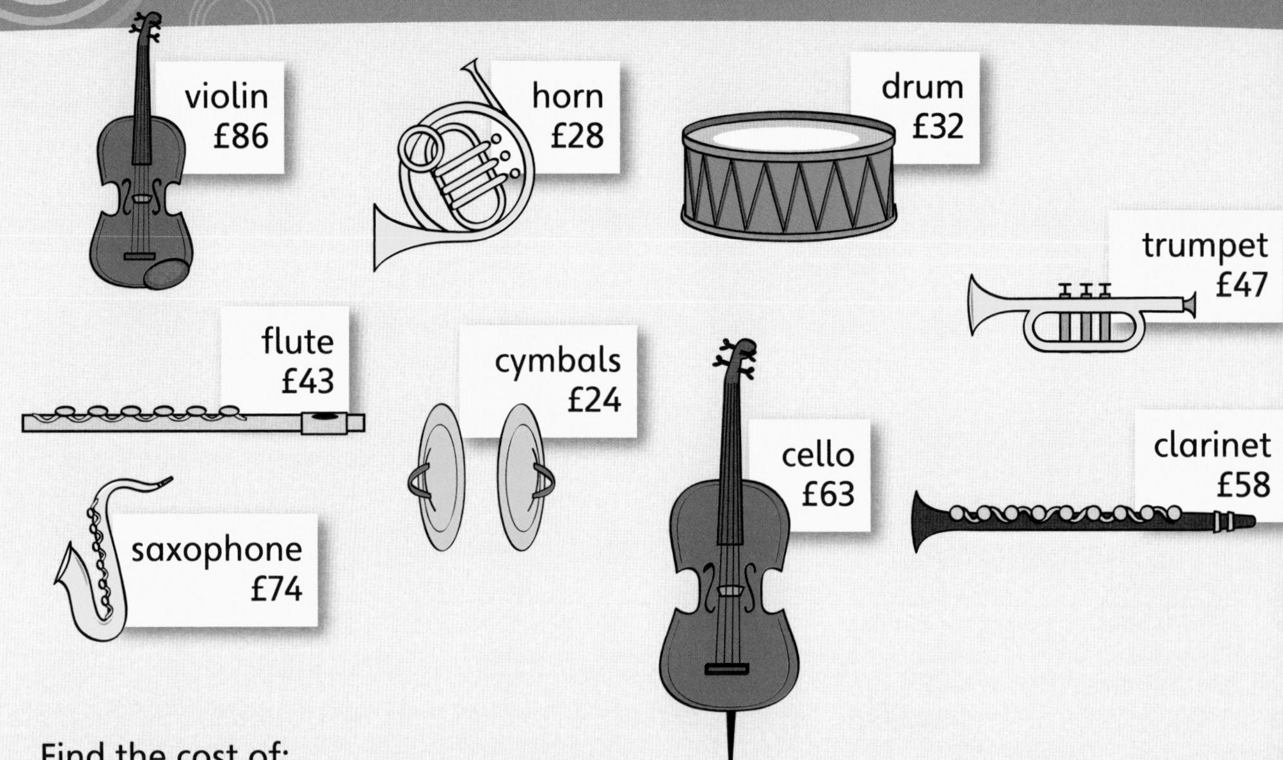

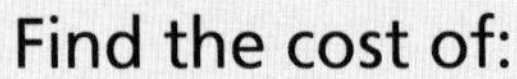

Find the cost of:

1 six violins
2 three drums
3 four trumpets
4 two flutes
5 six clarinets
6 three cellos
7 seven saxophones
8 four horns
9 two cymbals

1.				
		8 6		
	×	6		
		4 8 0		6 × 8 0
		3 6		6 × 6
	£	5 1 6		

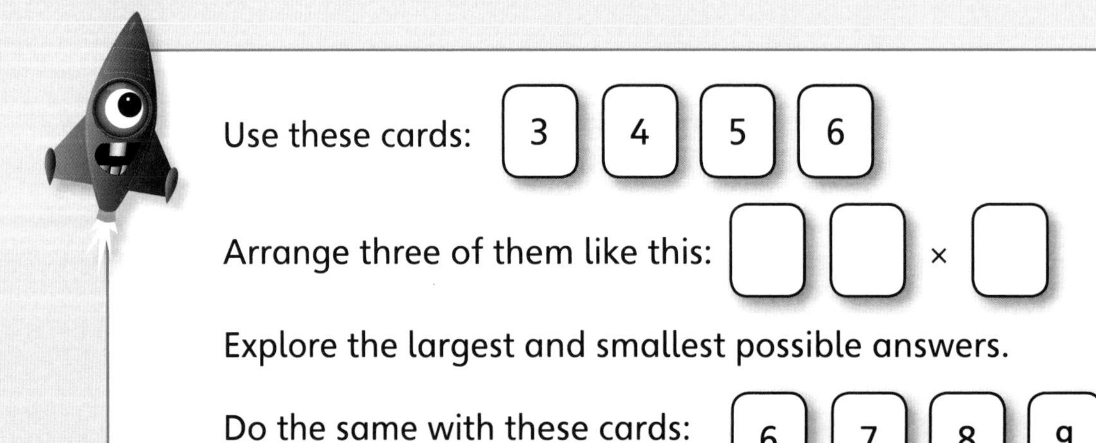

Use these cards: 3 4 5 6

Arrange three of them like this: ☐ ☐ × ☐

Explore the largest and smallest possible answers.

Do the same with these cards: 6 7 8 9

Multiplying

Copy and complete these multiplications. Write an estimate first.

1 (900)

```
  314
×   3
_____
        3 × 300
        3 × 10
        3 × 4
_____

_____
```

2 (800)

```
  186
×   4
_____
        4 × 100
        4 × 80
        4 × 6
_____

_____
```

```
1. (900)
   314
 ×   3
   900   3 × 300
    30   3 × 10
    12   3 × 4
   942
```

3 786 × 2 =	**4** 274 × 5 =	**5** 327 × 6 =	**6** 643 × 6 =
7 512 × 3 =	**8** 487 × 4 =	**9** 356 × 2 =	**10** 274 × 4 =
11 563 × 2 =	**12** 623 × 2 =	**13** 335 × 6 =	**14** 422 × 3 =

Find the cost of these holidays for a: 4 people, b: 5 people, c: 7 people.

15

£346 each

16

£543 each

17

£236 each

18

£318 each

19

£274 each

20

£437 each

Put digits in the boxes to make the calculation correct.
How many different ways can you find?

□□□ × □ = 648

Multiplying

Copy and complete these multiplications. Write an estimate first.

1 (1500)

```
  528
×   3
-----
 1500
   60
   24
-----

-----
```

2 (2000)

```
  464
×   4
-----
 1600
  240
-----

-----
```

3 (4200)

```
  732
×   6
-----
 4200

-----

-----
```

Now try these.

4 326 × 4

5 458 × 3

6 724 × 6

7 562 × 4

8 395 × 5

9 643 × 4

10 527 × 7

11 741 × 3

Doughnuts are sold in boxes of four. Calculate how many doughnuts were sold each week.

12 Week 1
426 boxes

13 Week 2
148 boxes

14 Week 3
276 boxes

15 Week 4
325 boxes

16 Week 5
542 boxes

17 Week 6
441 boxes

If a doughnut costs 5p to make, and a box of four doughnuts is sold for 45p, investigate how much profit the company makes each week.

I can split up a 3-digit number to help me multiply it

Multiplying

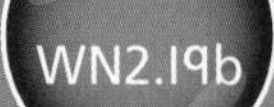

Complete the multiplications. Use the shorter method if you can.

1. 276
× 3
600 3 × 200
210 3 × 70
18 3 × 6
828

1 276 × 3

2 436 × 4

3 147 × 6

4 382 × 6

5 247 × 5

6 523 × 3

7 193 × 6

8 318 × 7

9 What is the difference between the cost of 3 season tickets at £425 each for City and 4 season tickets at £316 each for Rovers?

10 Jason has won a prize. He can choose between £178 each month for 6 months or £235 each month for 4 months. Which should he choose to get the most money?

Use the digit cards    

Make a 3-digit number and a 1-digit number, and multiply them together.

What is the largest answer?
What is the smallest answer?

How many different answers can you make between 2000 and 3000?

I can multiply a 3-digit number by a 1-digit number

Multiplying

Work out how much the coach company will receive on each trip.

1.	30	2
20	600	40
7	210	14

$$\begin{array}{r} 640 \\ +\,224 \\ \hline \pounds 864 \end{array}$$

1 Mystery tour £32

27 people

2 Sea and sand £42

18 people

3 Lakes and mountains £56

33 people

4 Famous gardens £33

32 people

5 City of London £63

41 people

6 Ancient castles £29

24 people

7 National parks £28

38 people

8 Birdwatching £42

26 people

9 Theme park £34

54 people

True or false?

10 $38 \times 23 = 23 \times 38$

11 $41 \times 25 = (41 \times 20) + (41 \times 5)$

12 $26 \times 30 < 36 \times 20$

13 $36 \times 27 > 37 \times 26$

I can multiply a 2-digit number by a 2-digit number

Multiplying

Calculate the total cost of these flights.

1.	200	30	8
20	4000	600	160
4	800	120	32

$$\begin{array}{r} 4760 \\ +\ \ 952 \\ \hline £5712 \end{array}$$

1 Vienna
tickets £238
party of 24

2 Barcelona
tickets £149
party of 28

3 Paris
tickets £117
party of 23

4 Nice
tickets £137
party of 18

5 Berlin
tickets £246
party of 25

6 Krakow
tickets £324
party of 19

Investigate the total cost for your whole class and your teacher to go on each flight.

Choose how to solve the calculations.

7 427 × 21 =

8 316 × 32 =

9 235 × 43 =

10 547 × 54 =

11 189 × 26 =

12 237 × 33 =

13 347 × 48 =

14 526 × 29 =

15 637 × 36 =

16 483 × 23 =

17 615 × 44 =

18 384 × 21 =

Multiplying

Copy and complete these multiplications.

1 274 × 23

	200	70	4
20	4000	1400	80
3	600		

5480
+ ...

2 156 × 32

	100	50	6
30	3000	1500	
2			

...
+ ...

3

	300	20	6
10	3000		
8			

4

	400	30	8
20			
6			

5

	200	60	7
30			
4			

Write the total miles each aeroplane flies.

6 London to Paris
213 miles
14 trips

7 Paris to Madrid
652 miles
23 trips

8 Vienna to Dublin
821 miles
32 trips

9 Rome to Berlin
734 miles
26 trips

10 Milan to Amsterdam
513 miles
19 trips

11 Barcelona to Zurich
525 miles
24 trips

Investigate approximately how many trips each aeroplane needs to make to cover 5000 miles.

Multiplying

Write an estimate by rounding each number to the nearest 10.

1. 20 × 30 = 600

1 23 × 31 **2** 42 × 29 **3** 38 × 22 **4** 47 × 18

Copy and complete. Write your estimate first.

5

6

7

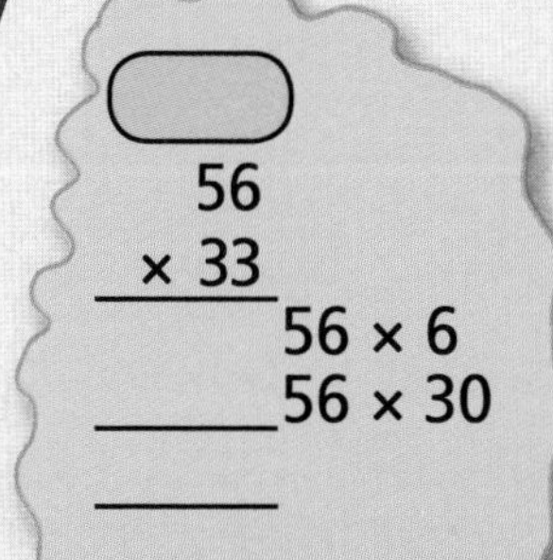

8 53 × 16

9 47 × 22

10 39 × 33

11 52 × 19

12 Amy has 1000 sweets. She gives 24 sweets to each child in her class. There are 31 children. How many sweets are left?

13 Janine is 34 years old. How many more weeks until she has lived for 2000 weeks?

14 Craig has 48 pieces of pipe, each 36 cm long. He joins them together. How long is the pipeline? There are 100 cm in a metre. How much longer to reach 20 m?

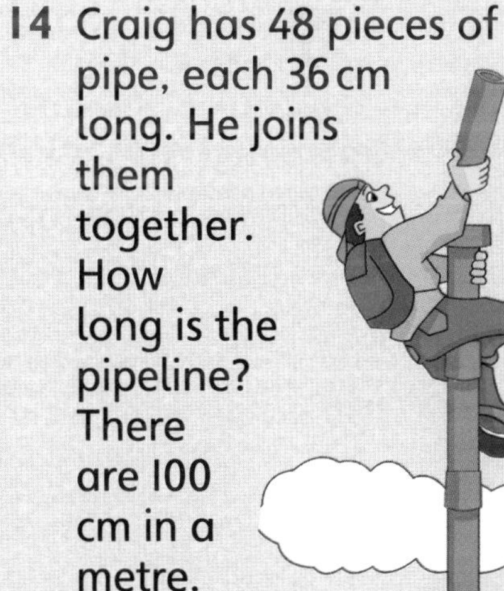

Write your own word problem using multiplication of 2-digit numbers.

I can multiply a 2-digit number by a 2-digit number

Multiplying

Complete these multiplications. Estimate first!

1 $\begin{array}{r} 46 \\ \times\ 27 \\ \hline \end{array}$

2 $\begin{array}{r} 28 \\ \times\ 34 \\ \hline \end{array}$

3 $\begin{array}{r} 56 \\ \times\ 29 \\ \hline \end{array}$

4 $\begin{array}{r} 52 \\ \times\ 43 \\ \hline \end{array}$

5 $\begin{array}{r} 28 \\ \times\ 37 \\ \hline \end{array}$

6 $\begin{array}{r} 53 \\ \times\ 17 \\ \hline \end{array}$

7 $\begin{array}{r} 64 \\ \times\ 27 \\ \hline \end{array}$

8 $\begin{array}{r} 63 \\ \times\ 32 \\ \hline \end{array}$

Write the area of these courts.

9

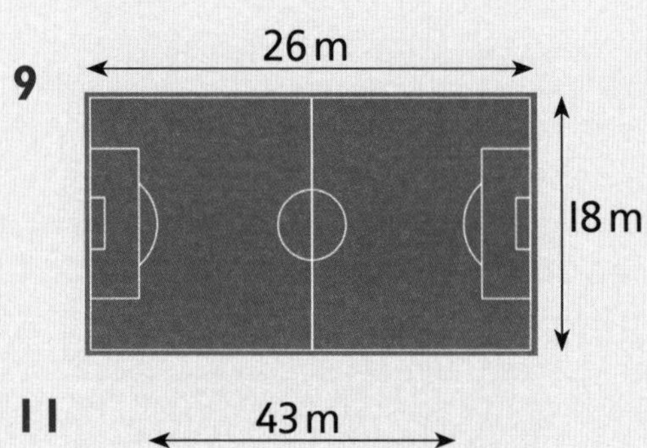

10

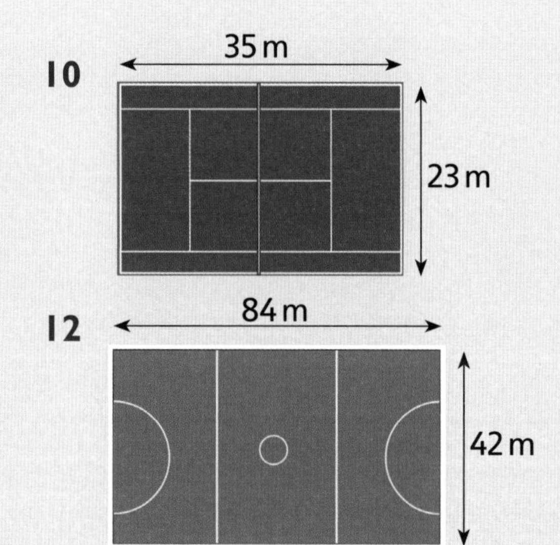

11

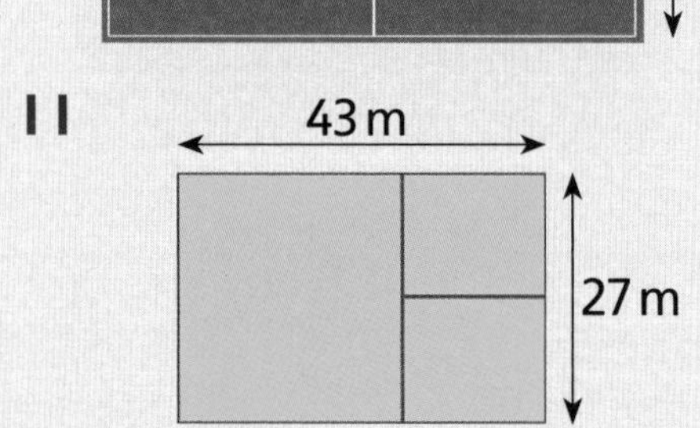

12

Use these place-value cards to make two 2-digit numbers.

20 30 40

3 4 7

Multiply them together.
Can you make a product over 2000?
Can you make a product that is a multiple of 3?

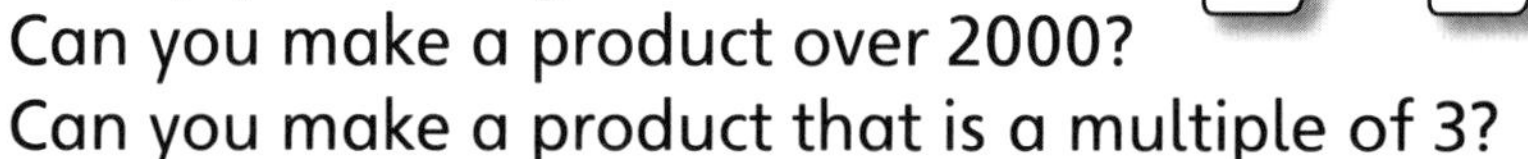

I can multiply a 2-digit number by a 2-digit number

Dividing using 10s

Do the divisions by chunking in 10s. You can use a number line to help you.

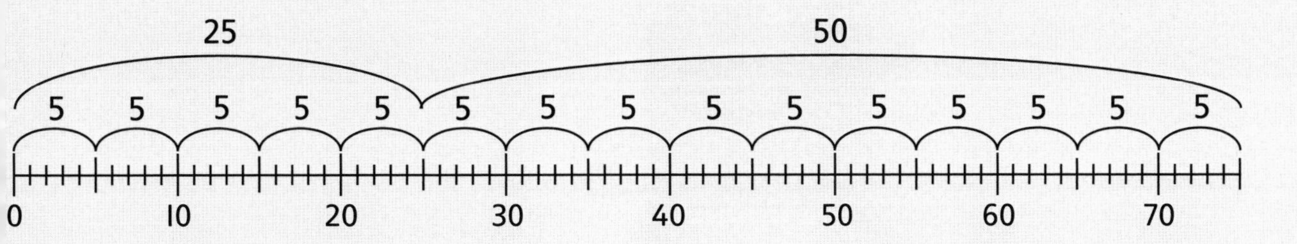

1 75 ÷ 5 =

```
          75
        - 50   (10) × 5
          25
        - 25   (5) × 5
remainder  0
```

75 ÷ 5 = 15 r 0

2 49 ÷ 3 =

```
          49
        - 30   (10) × 3
          19
        - 18   (6) × 3
remainder  1
```

49 ÷ 3 = 16 r 1

3 52 ÷ 4 = **4** 42 ÷ 3 = **5** 56 ÷ 4 = **6** 85 ÷ 5 =

7 42 ÷ 2 = **8** 39 ÷ 3 = **9** 96 ÷ 3 = **10** 79 ÷ 4 =

Lucas uses doubling and halving to work out mentally that 64 ÷ 4 = 16. Explain how he might have done this.

11 Jan shares 72 football cards between 6 friends. How many cards has he given to each friend?

12 Miss Damico is taking her class to a Musical. She has collected £84 from the class. Tickets cost £6 each. How many children are going?

I can divide a big number by taking away chunks

Do these divisions.

1.	96 ÷ 6	
	96	
	− 60	(10) × 6
	36	
	− 36	(6) × 6
	0	
96 ÷ 6 = 16		

1

÷6

96 ÷ 6

2

85 ÷ 5

3

78 ÷ 6

4

84 ÷ 7

5

96 ÷ 8

6

54 ÷ 3

7

÷4

76 ÷ 4

8

92 ÷ 4

9

64 ÷ 4

Which numbers between 80 and 100 can be divided exactly by 8? What other numbers can they be divided by?

Find the answers.

10 95 ÷ 5 =

11 57 ÷ 3 =

12 69 ÷ 3 =

13 75 ÷ 3 =

14 56 ÷ 2 =

15 92 ÷ 4 =

I can divide a big number by taking away chunks

Remainders

Do these divisions.

1. 75 ÷ 4
   75
   −60 (15) × 4
   15
   −12 (3) × 4
   3
   75 ÷ 4 = 18 r3

1 75 ÷ 4 =

2 39 ÷ 2 =

3 47 ÷ 3 =

4 58 ÷ 5 =

5 97 ÷ 6 =

6 54 ÷ 5 =

7 37 ÷ 7 =

8 67 ÷ 3 =

9 74 ÷ 3 =

10 Johan is planting tulips in 12 rows. He has 98 tulips. How many are in each row?

11 87 ÷ 5 =

12 55 ÷ 3 =

13 79 ÷ 4 =

14 73 ÷ 3 =

15 58 ÷ 8 =

16 86 ÷ 4 =

'Divisible by' means 'divides exactly with no remainder'. Investigate and complete these statements.

* A number is divisible by 2 when the number is even.
* A number is divisible by 5 when ____.
* A number is divisible by 10 when ____.
* A number is divisible by 3 when ____.
* A number is divisible by 9 when ____.

Dividing

Copy and complete these divisions.

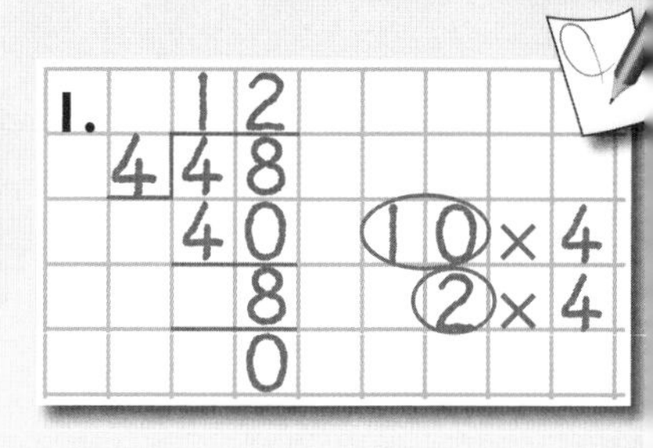

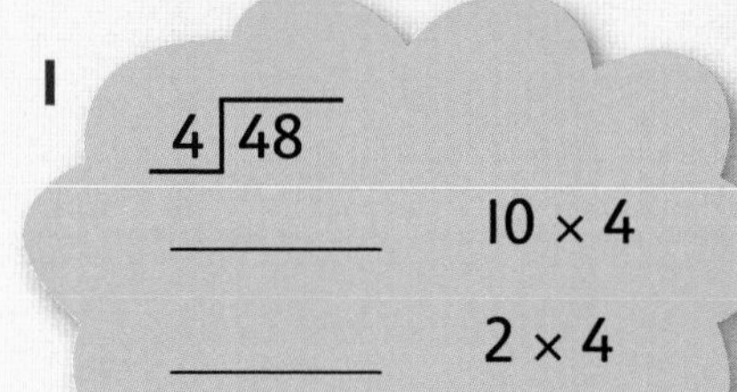

2 5|85 □ × 5 □ × 5

3 3|51 □ × 3 □ × 3

4 8|96 □ × 8 □ × 8

Each freezer is packed with ready meals.
Cara needs to eat 3 meals a day.
For how many days could Cara survive?

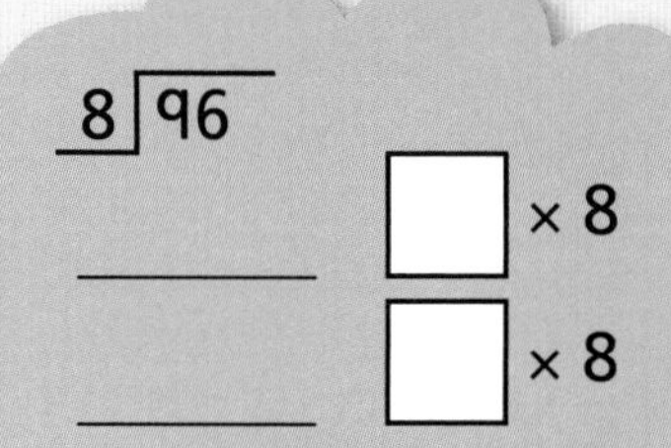

5

48 ready meals

6

54 ready meals

7

57 ready meals

For each freezer, how many days could Big Colin survive eating 4 meals a day?

I can divide a 2-digit number by a 1-digit number

Dividing

Kirsty thinks it will take her 4 days to read a book with 33 pages if she reads 8 pages each day. Hannah thinks it will take 5 days. Who is right? Explain your answer.

How many days will it take to finish each book?

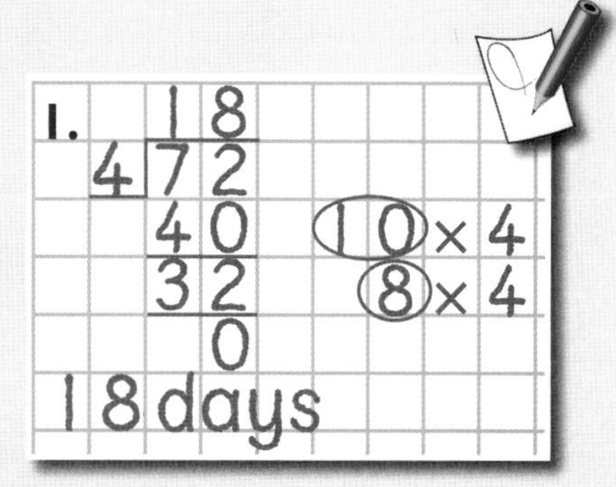

1

72 pages
4 pages a day

2

51 pages
3 pages a day

3

84 pages
6 pages a day

4

98 pages
4 pages a day

5

85 pages
3 pages a day

6 The people in questions 1 to 4 all agree to read 8 pages a day. How long will each person take over their book? Can you find a quick way to work out how long Person 1 will take? And Person 4?

I can divide a 2-digit number by a 1-digit number and find the remainder if there is one

Copy and complete.

1

```
     3 4  r
 4 | 1 3 8
     1 2 0  (30) × 4
       1 8
     _____  ... × 4
```

2

```
            r
 5 | 1 9 6
     1 5 0  (30) × 5
     _____  ... × 5
```

3

```
            r
 5 | 2 2 3
     _____  (40) × 5
     _____  ... × 5
```

4 $6\overline{)191}$ **5** $4\overline{)173}$ **6** $3\overline{)221}$

7 $5\overline{)316}$ **8** $4\overline{)190}$ **9** $6\overline{)487}$

10 $4\overline{)363}$ **11** $3\overline{)284}$ **12** $5\overline{)367}$

Write how many weeks it takes to save:

13 £165 at £3 a week

14 £233 at £4 a week

15 £191 at £7 a week

16 £346 at £5 a week

17 £213 at £6 a week

18 £471 at £8 a week

Investigate how many weeks it takes to save £500 at each different rate.

I can divide a 3-digit number by a 1-digit number and find the remainder if there is one

Dividing

Crackers are packed in boxes. Find how many boxes. Write an estimate first.

1. 140

$$\begin{array}{rl} 139 \text{ r } 1 & \\ 4\overline{)557} & \\ 400 & 100 \times 4 \\ \hline 157 & \\ 120 & 30 \times 4 \\ \hline 37 & \\ 36 & 9 \times 4 \\ \hline 1 & \end{array}$$

1 557 crackers

boxes of 4

2 613 crackers

boxes of 5

3 513 crackers

boxes of 3

4 728 crackers

boxes of 6

5 873 crackers

boxes of 7

6 922 crackers
boxes of 6

7 742 crackers
boxes of 3

8 937 crackers
boxes of 7

9 925 crackers
boxes of 7

10 496 crackers
boxes of 3

11 870 crackers
boxes of 3

One eighth of all crackers don't make a bang. How many of the total number of crackers are faulty?

12. 200

$$\begin{array}{rl} 228 \text{ r } 1 & \\ 4\overline{)913} & \\ 800 & 200 \times 4 \\ \hline 113 & \\ 80 & 20 \times 4 \\ \hline 33 & \\ 32 & 8 \times 4 \\ \hline 1 & \end{array}$$

Complete these divisions. Estimate first.

12 $4\overline{)913}$ **13** $7\overline{)854}$ **14** $3\overline{)726}$

15 $5\overline{)681}$ **16** $2\overline{)739}$ **17** $6\overline{)904}$

18 $3\overline{)858}$ **19** $4\overline{)703}$ **20** $2\overline{)937}$

I can divide a 3-digit number by a 1-digit number and find the remainder if there is one

Dividing

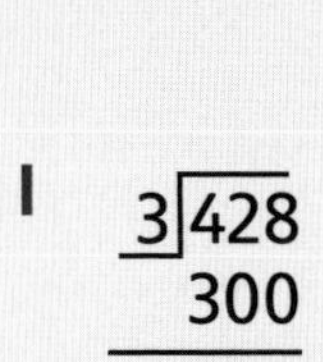

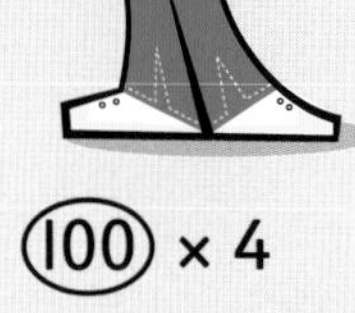

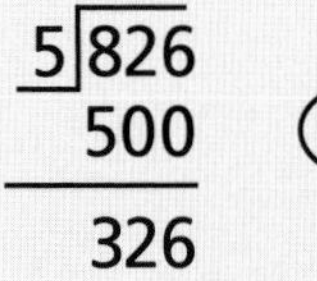

Copy and complete these divisions.

1 $3\overline{)428}$
300 (100) × 3
128
120 (40) × 3
8
(2) × 3

2 $4\overline{)573}$
400 (100) × 4
173
(40) × 4

3 $5\overline{)826}$
500 (100) × 5
326

4 $4\overline{)464}$ **5** $5\overline{)578}$ **6** $3\overline{)369}$ **7** $5\overline{)106}$

How many newspapers are there in each delivery?

8 
432 papers
3 deliveries

9 
517 papers
4 deliveries

10
643 papers
5 deliveries

11 
974 papers
5 deliveries

12 
588 papers
3 deliveries

13 
724 papers
4 deliveries

14
537 papers
3 deliveries

15 
821 papers
5 deliveries

16 
627 papers
4 deliveries

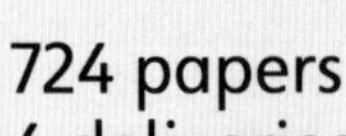

A newsagent has four paper girls and 636 papers to deliver. If one paper girl is ill, how many extra papers does each one deliver?

I can divide a 3-digit number by a 1-digit number and find the remainder if there is one

Dividing

Copy and complete using remainders.

1 $489 \div 20 =$

2 $574 \div 30 =$

3 $957 \div 50 =$

4 $849 \div 70 =$

5 $508 \div 20 =$

6 $714 \div 30 =$

7 $438 \div 50 =$

8 $258 \div 70 =$

9 $645 \div 40 =$

Hilary is planning a new kitchen. Her wall is 579 cm wide. In the shop, cupboards are 60 cm, 40 cm or 30 cm wide.

How many cupboards can she have, and how much space would be left, if she buys only 60 cm cupboards?

How many cupboards can she have, and how much space would be left, if she buys only 40 cm cupboards?

And if she buys only 30 cm cupboards?

Suggest a way she can fill her wall with cupboards of all three sizes.

10 $396 \div 40 =$

11 $904 \div 80 =$

12 $857 \div 30 =$

13 $478 \div 80 =$

14 $709 \div 40 =$

15 $1009 \div 20 =$

I can divide a 3-digit number by a multiple of 10 and find the remainder

Dividing

Copy and complete using remainders.

1. 28 r7
$30\overline{)847}$

1 $30\overline{)847}$ 2 $40\overline{)903}$ 3 $30\overline{)890}$ 4 $912 \div 30 =$

5 $40\overline{)607}$ 6 $350 \div 60 =$ 7 $70\overline{)904}$ 8 $70\overline{)935}$

9 $893 \div 20 =$ 10 $20\overline{)289}$ 11 $30\overline{)752}$ 12 $20\overline{)989}$

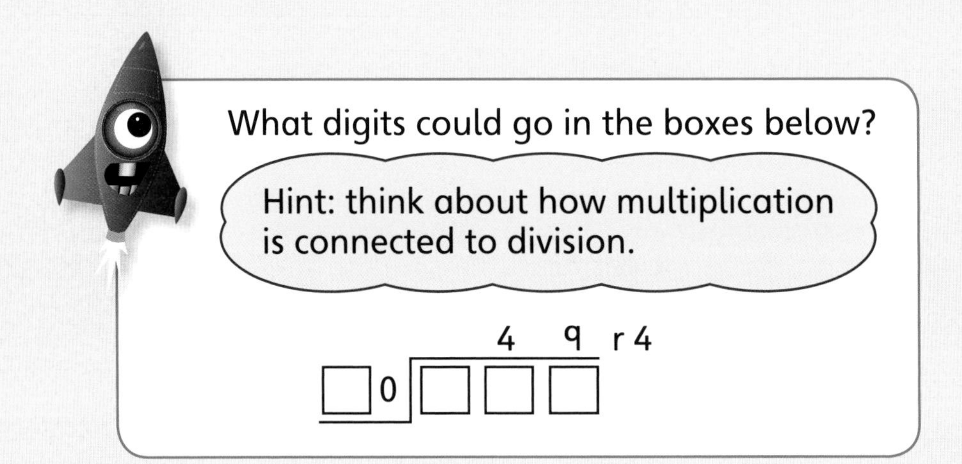

13 A lorry is 20 m long. How many similar lorries could fit on a street that is 584 m long?

14 Stuart is going to swim a mile. The swimming pool is 50 m long. One mile is about 1609 m. How many lengths will Stuart need to swim?

I can divide a 3-digit number by a multiple of 10 and find the remainder

Dividing

Copy and complete.

1 $2\overline{)33}$

2 $41 \div 2 =$

3 $2\overline{)57}$

4 $47 \div 3 =$

5 $5\overline{)69}$

6 $71 \div 6 =$

7 $8\overline{)94}$

8 $88 \div 6 =$

9 $8\overline{)93}$

10 $97 \div 9 =$

11 $7\overline{)94}$

12 $97 \div 6 =$

Selma keeps hens. She puts eggs in boxes of 6.
How many boxes can she fill each day?

13 Sunday 73 eggs

14 Monday 79 eggs

15 Tuesday 65 eggs

16 Wednesday 70 eggs

17 Thursday 74 eggs

18 Friday 71 eggs

19 Snow White won £75 worth of book tokens. She shared them equally among the seven dwarves and kept what was left for herself. How much did each get?

Louise has saved £122 for her holiday. She had £8 in her bank to start with and has saved £6 every week. How many weeks has she been saving?

Dividing

Write an estimate for the cost of each call.

1. $80p \div 4 = 20p$

1 83p for 4 calls

2 67p for 3 calls

3 97p for 5 calls

4 72p for 6 calls

5 91p for 4 calls

6 93p for 7 calls

7 107p for 5 calls

8 125p for 6 calls

9 157p for 8 calls

Write some divisions that will have an estimated answer of 20.

Copy and complete.

10 $63 \div 4 =$

11 $55 \div 3 =$

12 $87 \div 5 =$

13 $75 \div 4 =$

14 $55 \div 4 =$

15 $91 \div 6 =$

16 $96 \div 8 =$

17 $59 \div 3 =$

18 $66 \div 5 =$

I can solve problems about dividing 2-digit and 3-digit numbers

Dividing

Write how long it takes to finish reading these books.

1. 25 r 2   5 | 12²7   or

```
   25 r2
  |127
  -100
    27
    25
     2
```

rounds to 26 days

1 127 pages

5 pages a day

2 197 pages

4 pages a day

3 188 pages

3 pages a day

4 113 pages

6 pages a day

5 154 pages

9 pages a day

6 173 pages
7 pages a day

7 129 pages
5 pages a day

8 185 pages
8 pages a day

9 203 pages
6 pages a day

10 187 pages
4 pages a day

11 206 pages
8 pages a day

Find some long books. Work out how long it will take to read them at 4 pages per day.

Complete these divisions.

12 $177 \div 3 =$ **13** $113 \div 4 =$ **14** $123 \div 5 =$ **15** $147 \div 6 =$ **16** $183 \div 7 =$

17 $107 \div 4 =$ **18** $121 \div 5 =$ **19** $203 \div 8 =$ **20** $164 \div 3 =$ **21** $139 \div 6 =$

I can solve problems about dividing 3-digit numbers

Dividing

Complete these divisions using sharing.

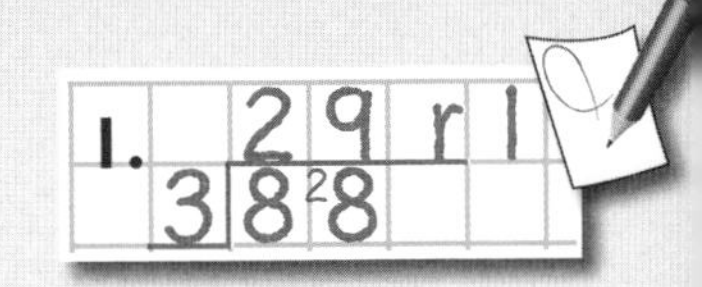

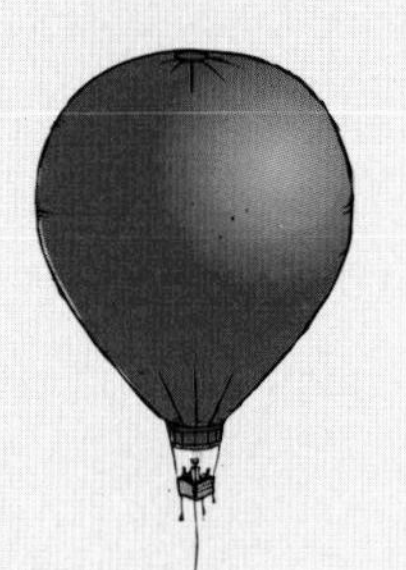

1 $88 \div 3 =$

2 $104 \div 6 =$

3 $167 \div 7 =$

4 $5\overline{)183}$

5 $8\overline{)241}$

6 $4\overline{)175}$

7 $119 \div 5 =$

8 $7\overline{)213}$

9 $204 \div 9 =$

10 $143 \div 8 =$

11 $163 \div 9 =$

12 $372 \div 7 =$

The answer to a division is 23 r 2.

If the number you divide by is a 1-digit number, what could the division be?

13 A taxi can take 4 passengers. How many taxis are needed for 71 children?

14 A number divided by 6 gives an answer of 5. What is the number?

15 Anna has 143 photos. An album holds 28 photos. How many albums does she need?

16 The CD rack can fit 7 CDs in each section. Shami has 165 CDs. How many sections will she fill?

I can use sharing to solve a division problem and make sure my answer is sensible

Dividing

Copy and complete using sharing.

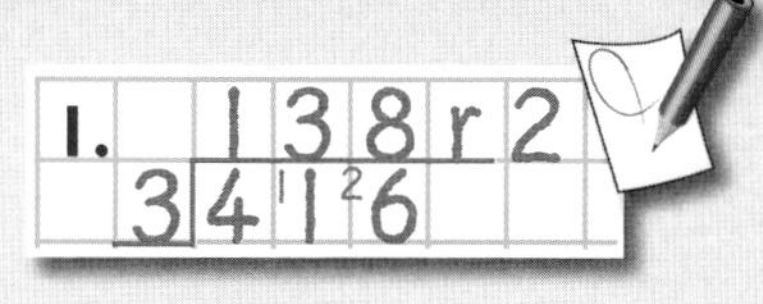

1 $3\overline{)416}$ **2** $5\overline{)627}$ **3** $4\overline{)739}$ **4** $726 \div 3 =$

5 $943 \div 4 =$ **6** $827 \div 3 =$ **7** $623 \div 5 =$ **8** $4\overline{)862}$

If there were the same number of visits to these websites each day, what would the average be?

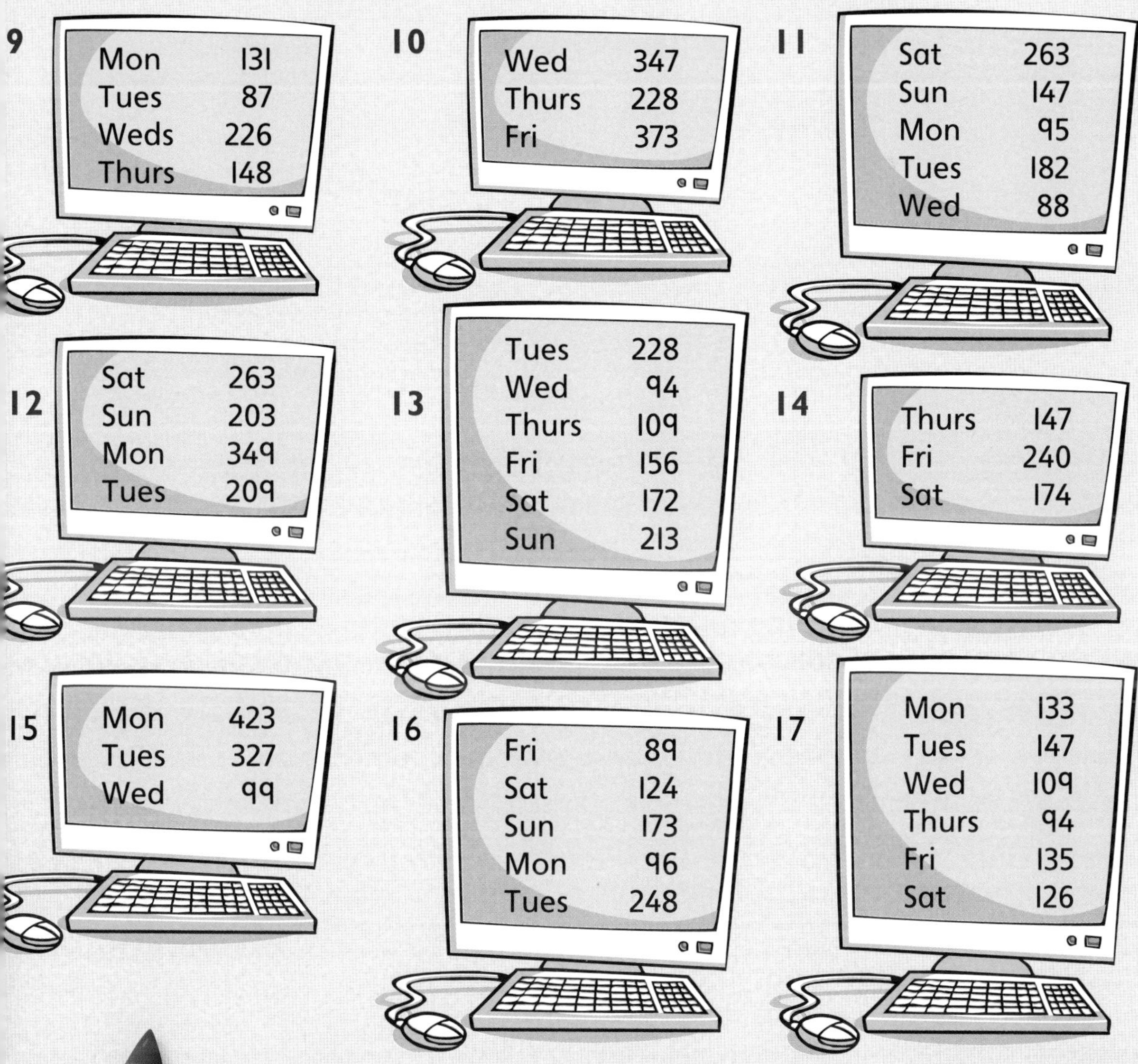

Investigate how many visits each website could expect in the month of June.

I can use sharing to solve a division problem and make sure my answer is sensible

How do we solve it?

1 Chang had 25p pocket money. Her dad doubled it. She spent 10p on a lolly. Her uncle doubled what she had left. How much does she have now?

2 Jane has saved £282 and Afram has saved £378. Find the total amount saved. How much more does Afram have than Jane?

3 Sean had collected 121 football cards. He gave Jason 68. How many did he have left?

4 The train travelled 328 miles on its journey. The last 162 miles were without heating! For how many miles was the heating on?

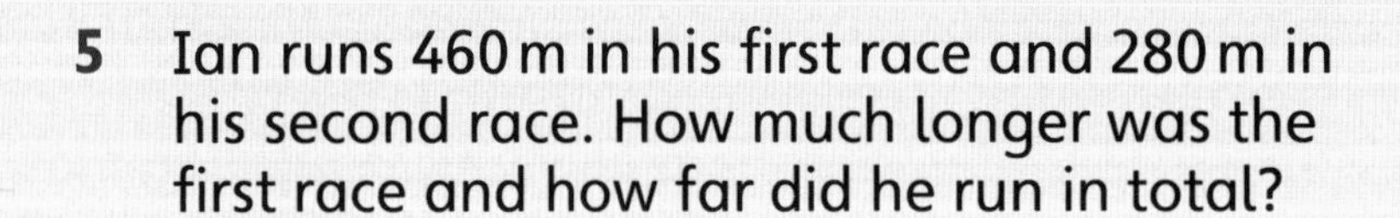

5 Ian runs 460 m in his first race and 280 m in his second race. How much longer was the first race and how far did he run in total?

A plane flies 3258 miles to London and 2469 miles to Athens. How far does it fly in total? What is the difference between the two distances?

I can read a problem and work out what it means, then choose the calculation I need to solve it

How do we solve it?

1 A rucksack weighs 350 g. Jamal adds a jacket weighing 280 g and boots weighing 490 g. What is the total weight?

2 Sarita weighed out 320 g of butter. When she turned her back, the cat ate 178 g. How much was left?

3 Ling had a satin ribbon that was 226 cm long. She cut off 152 cm. What length did she have left? How much less than half is that?

4 Tony had worked 328 days in 2009 and 287 days in 2010. How many days had he worked in all?

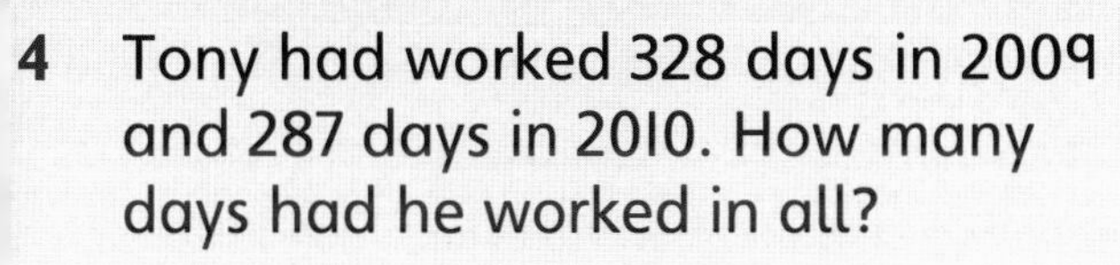

5 Tom the builder bought 5000 staples to fix the tiles on the roof. He used 4892. How many did he have left?

Anjilee has saved £360 towards a holiday costing £600. She thinks she can save £20 a week from now on. How long will it take her before she has enough money to pay for the holiday?

I can read a problem and work out what it means, then choose the calculation I need to solve it

How do we solve it?

1 At a concert there were 546 women and 484 men. How many people in all attended? If 128 people came after the interval, how many were there then?

2 Nick climbs 214 feet up a cliff. The top is 512 feet above the sea. He started at 28 feet above sea level. How far has he yet to climb? How far will he climb in total?

3 In one savings account Mr Rich had £4017. In another he had £3925. How much more did he have in the first account?

4 Sarita cycled 176 miles on her first cycling holiday and 158 miles on her second cycling holiday. How many miles did she cycle in total?

5 One aeroplane flew 4036 miles and a second one flew 3975 miles. How much further did the first plane fly?

The Chang family saved £1746 for the family holiday. It actually cost £2108. How much more did they have to pay?

I can read a problem and work out what it means, then choose the calculation I need to solve it

How do we solve it?

1 90 footballers arrive for a 5-a-side competition. How many teams can be made, and how many matches can be played in the first round?

2 Ling has 72 stickers, and can put 4 on each page in her album. How many pages can she fill?

3 Spring Water bottles are packed in sixes. 834 bottles are sent to the supermarket. How many packs of bottles is this?

4 Three friends go out for dinner, and the bill comes to £81. They share the bill and give a tip of £2·50 each. How much does each pay?

The three friends in question 4 go out for dinner again, and again share the bill. They each leave a tip of £3·50. Altogether they paid £102. How much did one meal cost?

I can read a problem and work out what it means, then choose the calculation I need to solve it

How do we solve it?

I Bina drinks lots of water. Last year she drank 937 pints of water. How many gallons is this? There are 8 pints in I gallon.

2 8692 people attended Rovers' last match. They all gave 5p to a charity. How much was collected? How much more would have been collected if they had given 8p instead?

3 It is I09 miles from Belfast to Dublin. Neelaksh makes I8 return trips in a year. How far has he travelled?

4 Last month the store sold I8 washing machines at £324 each and 2I dishwashers at £426 each. How much did they take altogether for these two items?

Next month, the store in question 4 plans to reduce the price on their washing machines to £300. They expect to sell 2I machines at this cheaper price.

If they manage this, how much more money will they make than last month?

I can read a problem and work out what it means, then choose the calculation I need to solve it

How do we solve it?

1 4732 people attended the theatre last week, each paying £8 for a ticket. The theatre company were hoping to receive takings of £40 000. How much more or less than this did they take?

2 Which is the greater prize: one-third of £828 or one-quarter of £948, and by how much?

3 Six years ago a new ship was built to travel round the Caribbean. The captain was told it would need an overhaul in 30 000 miles. Every year it has sailed 2768 miles. How many more miles can it do before it needs an overhaul?

4 Kate's number, divided by 3, gives an answer of 279 r 1. Paul's number, divided by 4, gives an answer of 236 r 3. What are the numbers? Which is larger and by how much?

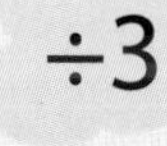

A 'Round the World' holiday costs £4286. Children travel half-price. What is the cost for a family of four adults and three children?

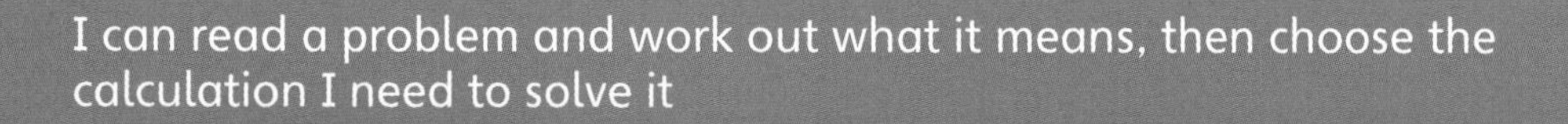

I can read a problem and work out what it means, then choose the calculation I need to solve it

Author Team: Peter Gorrie, Lynda Keith, Lynne McClure and Amy Sinclair
Consultant: Siobhán O'Doherty

Published by Pearson Education Limited, a company incorporated in England and Wales, having its registered office at Edinburgh Gate, Harlow, Essex, CM20 2JE. Registered company number: 872828

www.pearsonschools.co.uk

Heinemann is a registered trademark of Pearson Education Limited

First published 2012

15 14 13 12
10 9 8 7 6 5 4 3 2 1

British Library Cataloguing in Publication Data
A catalogue record for this book is available from the British Library

ISBN 978 0 435 07764 8

Typeset by Debbie Oatley @ room9design and revised by Mike Brain Graphic Design Limited, Oxford
Illustrations © Harcourt Education Limited 2006–2007, Pearson Education Limited 2010
Illustrated by Piers Baker, Fred Blunt, Emma Brownjohn, Tom Cole, Jonathan Edwards, Stephen Elford, Andy Hammond, John Haslam, Andrew Hennessey, Nigel Kitching, Sim Marriott, Q2A Media, Debbie Oatley, Andrew Painter, Tom Percival, Mark Ruffle, Anthony Rule, Eric Smith, Dale Sullivan and Gary Swift
Cover design by Pearson Education Limited
Cover illustration by Volker Beisler © Pearson Education Limited
Printed in the UK by Scotprint

Acknowledgements
The Publishers would like to thank the following for their help and advice:
Liam Monaghan
Hilary Keane
Stephen Walls

Every effort has been made to contact copyright holders of material reproduced in this book. Any omissions will be rectified in subsequent printings if notice is given to the publishers.